WILD BLUE

Also by Dan Bortolotti

Hope in Hell

DAN BORTOLOTTI

WILD
BLUE

A Natural History
of the World's
Largest Animal

Thomas Dunne Books

St. Martin's Press 🜨 New York

THOMAS DUNNE BOOKS.
An imprint of St. Martin's Press.

WILD BLUE. Copyright © 2008 by Dan Bortolotti. All rights reserved. Printed
in the United States of America. For information, address St. Martin's Press,
175 Fifth Avenue, New York, N.Y. 10010.

www.thomasdunnebooks.com
www.stmartins.com

Library of Congress Cataloging-in-Publication Data

Bortolotti, Dan.
    Wild blue : a natural history of the world's largest animal / Dan
Bortolotti.—1st U.S. ed.
        p. cm.
    Includes bibliographical references and index.
    ISBN-13: 978-0-312-38387-9
    ISBN-10: 0-312-38387-8
    1. Blue whale. I. Title.

QL737.C424 B67 2008
599.5'248—dc22

                                                            2008024933

First published in Canada by Thomas Allen Publishers,
a division of Thomas Allen & Son Limited

First U.S. Edition: October 2008

10 9 8 7 6 5 4 3 2 1

For my parents,
who taught me to think big

Contents

Preface   xi

1   Into the Blue   1
2   The Greatest War   21
3   Back to the Ice   47
4   About-Face   67
5   St. Lawrence Blues   93
6   West Coast Blues   121
7   Blues' Clues   143
8   Blue Notes   161
9   The 79-Foot Pygmy and Other Stories   193
10 Blue Genes   221
11 Why Size Matters   239
12 The Future Is Blue   259

    Author's Note   279
    Notes   283
    Sources   291
    Index   307

Preface

SQUINTING in the blazing July sunshine, I steered my rented car off U.S. Highway 101 onto Mission Street in Santa Barbara, California. I eventually made my way up a hilly road toward the Museum of Natural History, where I planned to occupy myself before catching a flight home from Los Angeles. The day before, I had spent the afternoon with biologist John Calambokidis as we launched his inflatable skiff from Ventura and motored out into the Santa Barbara Channel. About two hours into that boat ride, I saw my first blue whale. I had seen small whales before, but I was wholly unprepared for this meeting. The animal announced itself with a noisy plume of vapour that rose 20 or more feet in the air, and I was stunned by how long it took for the whale's back to roll through the water, its dorsal fin finally appearing after what seemed like a mile of mottled blue flesh. Over the next few hours, Calambokidis steered us alongside a dozen more blue whales, and each one was magnificent. By the time we returned to shore, I was sunburned, seasick and completely in awe of the magical animals I'd just had the privilege to see.

That experience was fresh in my mind as I walked toward the Santa Barbara museum, home to a 72-foot blue whale skeleton. And suddenly there it was, stretched out in front of the entrance, each rib curving way over my head, and with a skull so big I probably could have driven my rental car right inside it. As I walked up and down this immense exhibit to take it all in, a group of school kids arrived, perhaps 12 or 13 years old, and one of them caught sight of the bones. "Look, a dinosaur."

"No," another corrected him, "it's a whale."

A third boy gave the skeleton the briefest of glances before turning away. "Death by boredom," he pronounced.

I didn't get it. Children everywhere are fascinated by dinosaurs, which they know only from fossils and Hollywood special effects. Yet the skeleton of the largest animal species ever to inhabit our planet wasn't worth walking over to take a look. Never mind that the boy was probably unaware that living, breathing blue whales were at that moment feeding a couple dozen miles offshore. "Our enduring fascination with dinosaurs made me wonder," writes the wildlife biologist Douglas Chadwick, "if the world held no more whales, and they were known only from long-fossilized bones, would people show more interest in figuring out what the lives of the hugest animals of all time were like?"[1]

That trip to southern California was the first step in my journey to learn and share as much as I could about the lives of blue whales. The seed of my curiosity had been planted the year before, while I was editing a series of children's books on wildlife conservation. "The mysteries about blue whales," one of the authors had written, "greatly exceed the known facts."[2] It was a simple 10-word sentence, but I was taken aback. Why was so much about blue whales still unknown? How can an animal so large be so difficult to figure out? I eventually came to appreciate the difficulties involved in studying large whales, and to admire the scientists who were overcoming them. A month after my first visit to California, I returned and spent a week with Calambokidis's colleagues aboard the research vessel *John H. Martin* in Monterey Bay. I later journeyed to Quebec to spend a couple of weeks with Richard Sears, the biologist who pioneered the world's first long-term blue whale study in the Gulf of St. Lawrence and its estuary. I contacted scientists who work with blue whales in other areas—notably Australia, Chile, Indonesia and Antarctica—to learn how their work fit into the bigger picture.

While the contemporary research is exciting and important, one cannot tell the story of the blue whale without confronting the bloody history of whaling. There is no shortage of books from the whaling era that spin tales about the mightiest quarry of them all, though I quickly discovered that many of them are part fiction. I wanted to hear the stories first-hand—to talk to the men who had left their loved ones and voyaged to the Antarctic to join the whale hunt. I was fortunate to find several—including two who stood behind the harpoon cannon—who were willing to share their experiences.

Many of the people I spoke to told me that blue whales deserve a good book that weaves together the threads of history and science to tell the story of this enigmatic animal. These pages represent my best attempt to do that.

Dan Bortolotti
Aurora, Ontario
March 2008

WILD BLUE

ONE

INTO THE BLUE

T HE BLUE WHALE is at once the largest animal in the world and one of the most mysterious. It is the longest, heaviest and loudest living creature, and yet it can be remarkably inconspicuous. Only a tiny percentage of people will ever have the good fortune of seeing or hearing one in the wild. Humanity's attitude toward blue whales is also full of contradictions. Ancient cultures, so far as they knew about the mightiest of sea creatures, mythologized them. Modern humans, by contrast, chased them to the bottom of the globe, blasted them with exploding harpoons and rendered hundreds of thousands of them into soap and margarine. Today, the blue whale's place in our world has come full circle—once again held in awe, yet still little understood.

Perhaps the misunderstandings should not be a surprise. After all, the blue whale's dimensions are so gargantuan that humans have a hard time comprehending them. If someone told you that the sun's core temperature approaches 15 million degrees Celsius—which is true—would that figure have any real meaning? If another description increased the number by a few million degrees, would it seem any more implausible? The same issue seems to come up when describing blue whales: the animal's physical characteristics are exaggerated, often to the point of absurdity, yet we don't immediately notice. Many books and articles, for example, state that blue whales can be over 110 feet—a few stretch this to 115 feet—which almost certainly isn't true. The 2007 edition of *Guinness World Records* tells us that a blue whale's heartbeat can be heard up to 19 miles away, which is nonsense—no one has ever heard a blue whale's heartbeat, let alone from that distance. Others claim that blue whale calls are as loud as jet planes, heavy-metal concerts, even undersea earthquakes, and that blue whales routinely communicate across thousands of miles of ocean. Their diet, according to the venerable *National Geographic*, is up to 8 tons of krill per day. Think about that for a second: 16,000 pounds of food in 24 hours. All fantastic tales, to be sure, and all inaccurate, unproven, or at the very least highly misleading. It is as though the truth about blue whales wasn't astounding enough.

The truth *is* astounding, and it doesn't need to be stretched. Blue whales are indeed the largest animals ever to inhabit the earth. Once in a while, someone asserts that the mightiest dinosaurs were larger, but until there is a reliable method of determining a dinosaur's weight from a fossilized skeleton, this is pure speculation. Granted, a few dinosaurs may have been longer from tip to tail—although the longest scientifically measured blue whale was 98 feet, while the longest complete *Diplodocus* skeleton is less than 90 feet—but no prehistoric species has ever rivalled the blue whale's mass, which likely exceeded 200 tons in the burliest individuals. The vast majority of blues never reach these dimensions, of course—most today average about 70 feet and weigh perhaps 70 tons, while those in the Antarctic are typically about 80 feet and 100 tons. By comparison, only the heaviest African elephants weigh more than 6 tons. Roger Payne, the pioneering American scientist who has lived among whales for more than 40 years, vividly recalls the first blue he ever encountered: "Although by then I had probably seen as many whales as anyone alive, this creature made me feel I had never seen a whale at all."[1]

One might think that an animal so large would be thoroughly understood by scientists. But of all the popular misconceptions about the blue whale, this may be the most glaring. We live in a world where scientific discoveries are often so remarkable, and so far removed from our everyday experiences, that they give the impression that humans know almost everything. Astronomers have discovered more than 250 planets around distant stars since 1995. Biologists have sequenced almost the entire human genome. How hard can it be to observe an animal that is 70 or 80 feet long? Of course, it would involve many hours in boats and some fancy technology, but if we can clone a sheep, then surely we can understand the life history of a blue whale. The reality may come as an eye-opener.

Imagine a group of casual friends that you see each summer. If you go to their favourite restaurants when their preferred foods are in season, there's a good chance you'll run into them. When you do, however, they appear for just a few minutes each day, and what they do when you're not watching is a mystery. Indeed, sometimes you get the impression that you're missing out on whole aspects of their lives. You're not sure who they are dating. They speak to each other in a language you don't understand—in fact, their voices are so deep that you can't even hear their conversations. Every winter they disappear without telling you where they're going. Some never come back. Each year a few mothers return with new babies, but they won't tell you who the fathers are.

If your friends behaved like this, you would have some appreciation for what it is like to study blue whales. These animals do not give up their secrets easily.

There are about 80 species of whales, dolphins and porpoises—collectively called cetaceans—and these can be divided into two broad groupings. Toothed whales, or odontocetes, include all the dolphins and porpoises as well as sperm whales, orcas (killer whales), belugas and beaked whales. With the exception of the sperm whale, which can exceed 60 feet, most toothed whales are relatively small as whales go. The other group, known as baleen whales, or mysticetes, contains the true leviathans. These animals are distinguished by their unique feeding anatomy. Their mouths contain no teeth;

instead, they have a series of flexible, bristled plates called baleen, which they use to strain prey from the sea. The size and shape of these baleen plates varies radically—from less than 8 inches to more than 10 feet—as each species has adapted to its preferred type of prey, whether fish, copepods or tiny shrimplike crustaceans called krill. While some baleen whales feed on a combination of these, blues are the pickiest of eaters and prey almost exclusively on krill.

The two groups differ in other ways: odontocetes have a single blowhole, while mysticetes have two. Male toothed whales also tend to be bigger than females—in the case of sperm whales and orcas, the difference is extreme. In baleen whales, however, females are larger. Odontocetes also use a sophisticated form of sonar to locate their prey, sending out high-frequency clicks and whistles and listening for their echoes, much as bats do when foraging in the dark for insects. Baleen whales also have a complex repertoire of sounds, as we will explore in detail, but they do not use them for echolocating prey.

The baleen whale suborder is diverse, but with a few exceptions, it is made up of giants. Right whales and bowheads, the first to be targeted by European whalers more than a thousand years ago, reach 50 to 60 feet. The little-known pygmy right whale, found only in the southern hemisphere, is about a third as long. Grey whales get as large as 40 to 45 feet. All other baleen whales are lumped together in the family Balaenopteridae and are known as rorquals. Their most obvious identifying feature is the series of grooves that extend some two-thirds of the way along the ventral surface, or underside, of their body. These pleats allow the animal to expand its mouth into a cavernous pouch—somewhat like a pelican or a bullfrog—thereby taking in vast quantities of water and prey. This evolutionary adaptation, as it turns out, was key to these animals becoming the largest inhabitants of the ocean.

The origin of the word *rorqual* is unclear. The Norwegian *rør* means groove, tube or channel, while *hval* means whale. Many popular sources, reasonably enough, say the word means something along the lines of "grooved whale," a reference to the throat pleats.

Most dictionaries disagree, however, and some trace the first syllable to an older Norse word for "red." While it seems odd to describe rorquals this way—each species is some combination of black, grey, blue or white—there may be an explanation. The ventral pouch is light coloured, and when extended it can take on a pinkish appearance as it gets flushed with blood. Whatever the etymology, the rorquals include at least seven species. The oddball is the humpback whale (*Megaptera novaeangliae*)—its gigantic white-bottom pectoral fins, rounded body and the bumpy tubercles decorating its head are unique. The remaining rorquals are all classified in the same genus, *Balaenoptera*, and are similar in appearance, differing mainly in coloration and size. If a non-specialist were to look at silhouettes of all six with no indication of the scale, the species would be almost impossible to identify. In addition to the distinctive throat pleats, they have smooth and streamlined bodies and a small dorsal fin located close to the tail stock.

The smallest and most abundant rorqual is the minke, usually divided into two species: *Balaenoptera acutorostrata* in the northern hemisphere and *B. bonaerensis* in the Antarctic. Minkes are typically 23 feet in length and were once considered far too small to be worth hunting, though today hundreds are killed annually by Japanese and Norwegian whalers. The Bryde's whale (*B. edeni*), a tropical species reaching 49 feet, is the least understood rorqual, as it has never been hunted in great numbers. The slightly larger sei whale (*B. borealis*) is found in all oceans, usually at mid-latitudes, though it too was hunted in the Antarctic. The second-largest rorqual—and the second-largest animal on earth—is the fin whale (*B. physalus*), or finback, which is commonly 62 to 65 feet, though the largest females may exceed 85 feet. Finally, there is the grandest of them all—the blue whale (*B. musculus*), which averages 69 to 72 feet outside the Antarctic. In the Southern Ocean, whalers took individuals that exceeded 98 feet and may have weighed as much as 200 tons.

The rorqual species are so closely related that they occasionally interbreed. Researchers in the North Atlantic have seen individuals that resemble both blue and fin whales on many occasions, and

genetic analysis has confirmed at least four of them to be hybrids. One of these, killed in 1986 by whalers off Iceland, was even carrying a fetus—a significant discovery, since interspecies hybrids are typically sterile. Blue–humpback hybrids have also been reported, though not verified. Given this close relationship, the taxonomy is still being sorted out; there may turn out to be nine rorquals, if a couple of recent discoveries are confirmed.

Blue whales inhabit parts of all of the world's oceans, though humans generally encounter them only in their spring and summer feeding areas. Broadly speaking, there are about a dozen of these worldwide. In the North Atlantic, they include Quebec's Gulf of St. Lawrence, the Davis Strait, and the waters around Iceland and off the Azores. In the northeast Pacific, blues are found off southern California and on both sides of Mexico's Baja Peninsula. Blue whales are known to spend time in the western North Pacific, especially off Russia's Kamchatka Peninsula and the western Aleutian Islands, though no one has yet identified a popular feeding ground. They occur year-round in a region of the eastern tropical Pacific called the Costa Rica Dome. Moving south, blue whales have been encountered off Ecuador and Peru, though the only place off South America where they aggregate in relatively large numbers is in the fjords of southern Chile. They reliably visit waters off southern and west-ern Australia, Indonesia and Madagascar; in the north-central Indian Ocean they occur off eastern Sri Lanka and the Maldives. Finally, about 2,000 blue whales still roam the Southern Ocean, all that is left of a population that once numbered about 240,000.

Researchers today visit many of these feeding sites annually to conduct surveys and compile catalogues of individual blue whales. The first continuous blue whale study, however, was not launched until 1979, when the American biologist Richard Sears started his work in the Gulf of St. Lawrence, and later expanded his study area to other parts of the Atlantic. The thriving population of blue whales off California appeared rather suddenly in the mid-1980s, and those in southern Australia and Indonesia were poorly known to science until the late 1990s. The rediscovery of the feeding ground in southern Chile—it had formerly been known to whalers—was announced with some fanfare in 2003. As for the blues off Mada-gascar and Sri Lanka, they remain largely unstudied.

Blue whale research is conducted almost entirely in summer, and not simply because the weather is more amenable—there is also the nagging problem that no one is sure where the animals go in the fall and winter. Like the sun slipping below the horizon, these mas-sive creations simply vanish from our view until they are ready to reappear. No blue whale breeding ground has been found any-where in the world's oceans. For decades, biologists assumed that blues follow the same pattern as humpback and grey whales, who migrate from high-latitude feeding grounds in summer to tropical or subtropical waters to mate and calve. It now seems likely, how-ever, that there are no specific areas where large numbers of blue whale females gather to give birth. Instead, they seem to disperse at the end of the feeding season, using a number of different areas to mate and calve, most of them hundreds of miles from shore.

Exactly where the enormous population of Antarctic blues went in the winter was one of the great mysteries pondered by whalers. Females with fetuses near full term, as well as mothers with newborn calves, were sometimes observed in winter near shore stations in southwestern Africa, but almost no blues have been seen in this region for decades. The most vivid account is from Saldanha Bay, South Africa, in 1912, and involved a 95-foot mother:

> It had just given birth and was lying tired and still at the surface, when the whale-ship arrived and shot a harpoon into its back without it making any efforts to escape.
>   The calf was taken ashore. It was 7.03 metres long. The umbilical cord was still hanging on it, and the flukes were curled together. The most anterior baleen plates were just about to break through, while those at the back were already *ca.* 10 cm long. The ventral side was completely white in the grooved region, but behind had larger light-grey spots.[2]

Another blue whale birth—this one with a happier ending—occurred in 1946 when a very pregnant female became trapped in Trincomalee harbour, in eastern Sri Lanka. According to one report, a Royal Navy captain "approached the animal, attached a rope to her flukes, and proceeded to shift her into deep water."[3] The whale immediately returned to the harbour, however, where she gave birth the following day. The heroic captain is said to have towed the mother again to open water, where it was later seen swimming freely, although the fate of the calf is not mentioned.

If anyone witnessed the actual calving in Trincomalee harbour, there is no description of it. The closest anyone has come to beholding such an event was likely the gruesome spectacle that occurred on whaling ships when a pregnant female was killed and winched onto the deck, where the fetus sometimes slipped spontaneously from its mother's body. At some whaling stations, fetuses were routinely removed and measured, and in the late 1920s scientists were able to piece together a rough mating schedule for southern hemisphere blue whales: fetuses expelled in September averaged just over a foot in length, while those measured in December had grown to almost 7 feet. By April, fetal blue whales averaged 16 feet, still well shy of the typical birth length of about 23. The biologists extrapolated that the females visiting the Antarctic in the austral summer conceived between late June and late August, and that gestation was between 10 and 12 months. They also reasoned that if mothers nursed for seven months and then returned to their feeding grounds to recuperate before mating again, it was impossible for females to have more than one offspring every two years. Exactly when female blues reach sexual maturity is not known, though recent research suggests it happens at about age 10. And since multiple births in whales are extremely uncommon, each female produces no more than two, or perhaps three, calves in her first decade and a half. This exceptionally low reproductive rate—and humanity's ignorance of it—would have terrible consequences for the blue whale.

Just as no scientist has ever observed a birth, none has ever been a voyeur during a blue whale coupling. Yet despite our almost complete ignorance of the species' sex life—or perhaps because of it—accounts of the blue whale's phallic grandeur abound. According to some, their penises are 16 feet long, and they can expel 400 gallons of semen in a single ejaculation. These claims, disappointingly, are nothing but urban legends. An adult blue whale's penis is typically between 7 and 8 feet and the combined weight of the testicles—which were systematically measured during the whaling era—range from about 65 to 165 pounds. As in all whale and dolphin species, the male sexual organ is housed internally—swimming with a 7-foot penis exposed would be like driving a boat with the anchor hanging over the side—and is extruded through a genital slit when needed. Whale penises are tougher and more fibrous than those of other mammals, and it is believed that they use the elasticity of this tissue, rather than blood flow, to achieve an erection, but this process is not well understood. The world awaits a graduate student with the courage to study it.

The sexual organs and mammary glands of female blue whales are not conspicuous either. In fact, the vulva is not radically different in appearance from the genital slit in males, making it almost impossible for researchers to tell the boys from the girls during observations in the field. One sure way to identify a blue whale as female, however, is to spot one with a calf in tow—an adult seen close to a calf for any prolonged period is almost certainly its mother. Other adults, including the father, do not play a significant role during the seven months that the calf is nursing. Not that she needs the help—a lactating blue whale is a tremendous energy machine. Because whales must surface to breathe, they have evolved two adaptations that enable them to speed up the nursing process and minimize the time that mother and calf spend underwater. The first is breast milk with 35 to 50 percent fat—about ten times richer than cow's milk—which is designed to accelerate the calf's growth. In addition, like the pressurized hoses that quickly fill the fuel tanks of race cars at pit stops, the mother's muscular action actively discharges the milk into the calf's mouth during suckling. A fuel-injected calf can gain 40,000 pounds before being weaned, which works out to about 8 pounds an hour.

Other anatomical features of the blue whale are more mysterious. Blues are known to emit the most powerful vocalizations of any animals on earth, though many of their calls are at frequencies below the threshold of human hearing. How the animals generate such energetic sounds is completely unknown—no sound-producing organ or resonating membrane has been identified. Until someone figures out how to capture and study a live blue whale, this seems destined to remain unknowable.

Scientists do not even know how long blue whales live. Yet again, one needs to appreciate the challenge of answering this question. If an animal can never be kept in captivity, then scientists must somehow learn the year of birth and death of individuals in the wild, as they do with banded birds, or they must discover some body part than can indicate age, like the rings in a tree trunk. So far the oldest blue whale confirmed from direct observation is 38—it was first photographed in 1970 and sighted again in 2008. But scientists have been photo-identifying individual blue whales only since the 1980s, and older animals will undoubtedly turn up as long-term studies continue.

As for clues in the whales' anatomy, there are several, though each has its own set of uncertainties. The oldest method for aging blue whales involves examining the ovaries of sexually mature females for knobs of tissue called corpora albicantia, which are laid down each time the animal ovulates. If this happens about once every two and a half years—traditionally considered the maximum rate at which blue whales can bear calves—and if females begin ovulating at age 10, then six corpora would indicate an age of about 25, while eight would translate to 30 years. The classic study of blue whale ovaries in the 1930s found a small number with more than 30 corpora, which would make the whales older than 85, and a Japanese scientist in the 1970s reported an individual with 40 corpora, suggesting the whale was 110 years old. But this method has at least two shortcomings. First, if biologists are wrong about the rate of ovulation, or the age at which females become sexually mature, their age estimates could be way off. If females ovulate every year and a half, for example, then 30 corpora works out to age 55, not 85. The other unknown is whether blue whales eventually stop ovulating. If so, they might live for decades afterward, as any menopausal woman will be quick to point out, and this method would be unable to determine their maximum lifespan. It is quite possible, however, that they keep breeding until they drop—biologists have examined pregnant fin whales that were known to be older than 40.

Scientists tried a second method of estimating age in the 1950s, when they discovered that rorquals have a wax plug in the outer ear made up of alternating layers of light and dark. These ear plugs have been just as problematic, however. To begin with, it was not at first clear whether the whales added a light layer one year, followed by a dark one the next, or if their annual growth consisted of both layers together. By the 1980s it became clear that the latter was the case for fin whales, at least, meaning that early age estimates may have been off by a factor of two. (Whether this is also true for blues is unknown, though the two species are similar enough that it is likely.) There is also lingering uncertainty about whether the ear plug method is reliable for very young or very old individuals. That said, biologists have found blue whale ear plugs with at least 46 (presumably annual) layers, and one female with 33 layers was pregnant when she was examined.

Taken together, all this evidence means blue whales can certainly live to 37, can almost surely exceed 50, and may well be among the longest-lived animals in the ocean. If a maximum lifespan of 90 to 100 years is ever proven, it would not surprise many scientists. Some truly remarkable evidence for the longevity of baleen whales comes from a third technique for establishing age, which involves measuring the presence of a certain amino acid in the lens of the eye. (Again, this method has never been used on blues.) Scientists in the late 1990s examined the eyeballs of 48 bowhead whales and found four that were well over 100 years old—the oldest was reportedly 211. If this sounds hard to believe—and the 211-year-old result may well be inaccurate—corroborating evidence keeps showing up in whales killed by aboriginal hunters. At least seven bowheads hunted since 1981 have been cut open to reveal fragments of old harpoon tips, some believed to date back to the 19th century—in May 2007, a bowhead killed off Alaska contained fragments of an exploding harpoon that was patented in 1879. Other researchers have discovered that the narwhal, a small Arctic species, can live to 115 years old, though a comparable study on minke whales found a maximum age of 32. All of which means it is difficult to make generalizations. It would not be a stretch to suggest, however, that the entire era of hunting blue whales in the Antarctic—from 1904 until 1966—spanned a single generation. If an individual did manage to stay alive for these 60-some years, the feat would have been more remarkable than surviving a firing squad.

Which leads us to perhaps the most high-profile blue whale mystery of all: how many are left alive? Before getting to that question, it helps to know that the Antarctic population alone was between 200,000 and 300,000 just a century ago. During the first seven decades of the 20th century, whalers killed more than 330,000 blue whales in the Southern Ocean, and perhaps another 50,000 in other regions. By the early 1970s, only a few hundred remained in the Antarctic—as little as 0.1 percent of the population's original size. In other words, humans killed about 999 out of every 1,000 Antarctic blue whales in less than 70 years. While no one has made a scientifically valid estimate of the current global population, the best data suggest it is upwards of 10,000. In the Antarctic, they appear to have increased to more than 2,000—still less than 1 percent of their original number. Whales have a low natural mortality rate, and if humans leave them undisturbed they can, and do, recover. Indeed, this seems to be true of virtually all whale species. Whether the Southern Ocean will ever support hundreds of thousands of blues again, however, is an open question. The Antarctic blue whale is making a comeback, but it is a long climb out of hell.

———————

Dolphins and other small cetaceans appear in the mythology of many ancient cultures, and large whales—often indistinguishable from sea monsters—appear in stories as diverse as the Book of Jonah, Pliny's *Naturalis Historia*, the legend of St. Brendan, the *Arabian Nights Entertainments* and *Paradise Lost*. But only one culture's mythology includes the blue whale specifically. The ethnographer R.H. Barnes describes how the people of Lamalera, a village on the Indonesian island of Lembata, claim to have been brought from a nearby island on the back of a giant whale:

> An ancestor of the clan went to the shore at his original home to look for a small boat. There he saw a very large baleen whale. He asked the whale if it would take him. The whale agreed, so the man got a bamboo pole, like those they tie to lontar palms as ladders. He made a hole in the whale and stuck the ladder in it. . . . When the whale dived, he climbed the ladder to stay out of the water.[4]

Lamalera is one of the few surviving outposts of traditional whaling today, but the hunters target only sperm whales. Blues are the only large baleen whales seen near the island, and they are left unharmed. "There was a legend that the blue whale once saved our ancestors from drowning in the sea," a village leader told the *Jakarta Post* in 2007. "So we don't kill blue whales."[5]

Not until the late 17th century was the blue whale formally identified in Europe. British and Dutch whaling was in full swing during the 1600s, but blue whales were not among the catch, so naturalists had to wait for specimens to be delivered to their shores by the forces of nature. Toward the end of the century, the first recorded blue whale was served up in this fashion to Sir Robert Sibbald. In the 1680s, Sibbald was a founder of the Royal College of Physicians of Edinburgh, the Geographer Royal to King Charles II and the first professor of medicine at the university in Scotland's capital. Sibbald also published several writings on the flora and fauna of his native country, including a 1694 work entitled *Phalainologia Nova: or, observations on certain of the rarer whales recently stranded on the coast of Scotland*, which contains the first published description of a blue whale:

> In the month of September of the year 1692, on the south shore of the Firth of Forth, near the ancient fortress of Abercon, was cast up a male whale 78 feet long . . . It was believed that its girth exceeded 35 feet . . . In the upper jaw the whole palate was seen to be covered with black hairs, or rather bristles, which hung above the tongue, with which, at the sides, equally separated, appeared black, horny plates . . . No blowhole was present in this beast, but toward the forehead there were to be seen two large apertures approaching a pyramid in shape . . . The lateral fin was 10 feet long, 2½ feet broad where widest . . . The penis, which hung from the body not far from the navel, was 5 feet long, where thicker it was 4 feet in girth, and it gradually diminished to a very narrow extremity . . .

The tail, from that part in which it was divided into two flukes to the upper extremity, was 10 feet long; the distance between the two extremities of this (the flukes) was 18½ feet.[6]

Carolus Linnaeus relied on this account by Sibbald when he created his famous system of classifying animals and plants, making it the basis for one of the four whale species he named in the 1758 edition of his *Systema Naturae*. For more than a century afterwards, however, the animal that would later be called the blue whale remained obscure. While European naturalists wrote a succession of books and papers on the great whales, the vast majority had never actually seen one, and their descriptions often simply repeated— one might say plagiarized—the work of those before them. There were few opportunities for even the most conscientious natural historians to examine specimens, and even well into the 19th century illustrations of whales were hilariously inaccurate, complete with fountains of water spouting from the blowholes. In Henry William Dewhurst's *The Natural History of the Order Cetacea*, published in 1834, the drawing of the "broad-nosed whale"—an old name for the blue—wears a Hitlerian moustache. It is hard to know what inspired this artistic fancy—rorquals, especially young ones, do have a small number of short hairs on both the lower and upper jaw, but nothing so conspicuous. Perhaps the illustrator was influenced by the scientific term *mysticete*, which is from the Latin for "moustached whale"—though the reference is to the bristly baleen, not facial hair.

If Dewhurst had no first-hand knowledge of live rorquals, he did provide a detailed description of what may have been the first publicly exhibited blue whale skeleton. On November 4, 1827, he explained, some fishermen discovered an enormous dead rorqual in the North Sea and, with the combined efforts of three ships, managed to tow it to the harbour of Ostend, in what is now northern Belgium. "The appearance of a whale of such enormous dimensions created a great sensation," Dewhurst wrote, "inasmuch as those

La Baleine d'Ostende,
Visitée par l'Éléphant, la Giraffe les Osages et les Chinois.

The first blue whale to be publicly exhibited, the Ostend Whale toured Europe in the 1820s. Despite this fanciful illustration, only the skeleton was displayed.

which had formerly been stranded or captured on the coast of Flanders were of much smaller dimensions." He described it as "the largest animal that has ever been captured," at 95 feet and 249 tons.[7] Those measurements are certainly exaggerated: no northern hemisphere blue whale has been known to reach 95 feet, and 249 tons is wholly implausible. (How one would have weighed a large whale in a 19th century Flemish town is conveniently left to the imagination.) But whatever the whale's actual dimensions, there is no question that it was carefully dissected, and its bones were prepped for display. The exhibitors built a portable wooden pavilion to house the skeleton, and the curiosity became a travelling road show for at least seven years, with exhibitions in Holland, France and England. In 1856, the Ostend Whale, as it was known, was taken to Russia, where it still resides in the collection of the Zoological Museum in St. Petersburg.

It is easy to imagine the wonder of 19th-century Europeans as they beheld a display like the Ostend Whale. Indeed, itinerant exhibits like this continued for more than a century, since they provided the only opportunity for the public to lay eyes on a giant whale, albeit a dead one. Some impresarios, not content simply to mount the bones, even exhibited intact whales. In December 1880, a Massachusetts businessman named George Newton launched a tour of the midwestern United States featuring "The Prince of Whales," a 60-foot blue whale carcass. Newton was apparently able to keep the whale at least partially frozen as the Prince made its way by railroad to Chicago, Milwaukee, St. Louis, Cincinnati, Louisville, Columbus, Cleveland, Philadelphia and Buffalo. More than 100,000 people reportedly paid to see the whale—"at 25¢ for adults, 15¢ for children, orphans free"[8]—but when it arrived in Michigan in May 1881 the carcass was decaying too fast. Despite the valiant efforts of several Detroit butchers and an overmatched taxidermist, the Prince's reign was over. By the time P.T. Barnum arrived in America the following spring with a smaller though less smelly animal attraction named Jumbo the Elephant, the spectacle had been long forgotten.

In the late 1960s, the American Museum of Natural History in New York attempted to build a full-sized, realistic model so people could get some idea of what a blue whale would look like in life. The few museums with blue whales at that time tended to mount them in decidedly unnatural poses—the Smithsonian's original model, "with its flippers outstretched like stubby wings, did look a bit like a sausage coasting along in a low flight pattern."[9] A team of New York craftsmen worked for more than two years on the model, forced to guess at some anatomical features that were not well understood. The 94-foot whale, weighing some 21,000 pounds, was finally unveiled on February 26, 1969, to the amazement of 35,000 visitors, breaking the museum's attendance record. Ingeniously suspended without wires—it is attached to the roof trusses at a single, inconspicuous point—it appears to hover in mid-air, its form arching gracefully as if the animal is about to dive. (The model is even more impressive today, having undergone a makeover in 2003 to make its shape and colouring more accurate.) No doubt many of the New Yorkers who walked beneath the model that winter day in 1969 also visited the museum's famous *Tyrannosaurus* and *Brontosaurus* fossils. A few must have wondered whether the blue whale would soon be joining these dinosaurs in the pantheon of magnificent, extinct animals.

The largest whale species went by various names in the 19th century, most commonly the broad-nosed whale, Sibbald's rorqual and the sulphur-bottom. This last name, which lingered until the mid-1900s, derives from the thin film of yellowish diatoms (a type of algae) sometimes visible on its ventral surface. In *Moby-Dick*, published in 1851, Herman Melville used this moniker in his discussion of the classification of whales:

> BOOK I. (FOLIO), CHAPTER VI. (SULPHUR-BOTTOM).—Another retiring gentleman, with a brimstone belly, doubtless got by scraping along the Tartarian tiles in some of his profounder divings. He is seldom seen; at least I have never seen him except in the remoter southern seas, and then always at too great a distance to study his countenance. He is never chased; he would run away with rope-walks of line. Prodigies are told of him. Adieu, Sulphur Bottom! I can say nothing more that is true of ye, nor can the oldest Nantucketer.[10]

A few pages later, Melville includes another list of species:

> But there are a rabble of uncertain, fugitive, half-fabulous whales, which, as an American whaleman, I know by reputation, but not personally . . . The Bottle-Nose Whale; the Junk Whale; the Pudding-Headed Whale; the Cape Whale; the Leading Whale; the Cannon Whale; the Scragg Whale; the Coppered Whale; the Elephant Whale; the Iceberg Whale; the Quog Whale; the Blue Whale; etc.[11]

According to the *Oxford English Dictionary*, this is the first known usage of the term *blue whale*. However, from the context it is not clear whether Melville's blue whale is even meant to describe something real. Indeed, the whole chapter is written with irony and takes humorous digs at the scientific authors of the day. Melville's knowledge of rorquals would have been second- or third-hand, but even if he was aware of a species much larger than the one featured in his novel, he may have chosen to ignore it, as the scholar Lauriat Lane Jr. explained:

> Wittingly or unwittingly—the charge stands not proven— Melville had to slight the blue whale . . . Melville himself had sailed in the sperm whale fishery. To maim Captain Ahab, Melville's moral drama demanded a very large, toothed cetacean. Ecologically, erotically, and psychoanalytically the hunt had to be for the sperm whale. Thus, epically and allegorically, the hunt for the sperm whale had to be for the largest of all whales. And Melville, wrongly, made it so.[12]

The *OED*'s second citation for "blue whale" is from the 1888 edition of the *Encyclopaedia Britannica*, and is unequivocally the same animal for which we use the term today. But this came 37 years after *Moby-Dick*, and it is during those intervening years that the name was really coined. It originated in Norway (as *blåhval*) shortly after the harpoon cannon was perfected by Svend Føyn, the father of modern whaling. Føyn immediately turned his new invention on the species that had previously eluded human hunters, and in 1874 a Norwegian scientist examined some of these specimens for the first time:

> The whole ground color of the whale, seen at a distance, has a very distinctly bluish cast, and that in a more conspicuous man- ner than in any other whale with which I am acquainted. The name 'Blue whale,' bestowed on this species by Foyn, seems to

me very suitable, and I will therefore propose that it be adopted for the species as the Norwegian common name.[13]

In only thirty years or so, Føyn and his harpoon cannon would succeed in blasting his newly named whale to near extinction in the seas off northern Norway. By the beginning of the new century, however, expeditions to the Antarctic returned with news that the Southern Ocean held more blue whales than anyone could have imagined. Melville's "seldom seen" sulphur-bottom was about to become the target of an all-out assault.

TWO

THE GREATEST WAR

No human industry followed a more reckless, myopic pattern than whaling. From the beginning, whalers concentrated on one species, hunted it intensively until there were so few that it was uneconomical to continue, then moved on to a different species or another part of the ocean and did the same. Until the late 19th century, blue whales avoided being the next species crossed off the list because they were too fast and too heavy to haul in, and they lived too far out to sea for shore-based whalers to pursue them. But the Everest of whales remained the ultimate goal, and once technology allowed humans to overcome their handicaps, they made up for lost time. Hunting a species to near extinction in the era of open boats and hand harpoons could take hundreds of years, but modern whalers almost succeeded in emptying the oceans of blue whales in a few bloody decades.

Humans have killed whales since prehistoric times, though not in any systematic way at first—perhaps one ventured into a harbour and was forced onto the beach by men in boats, or hunters killing seals on the ice managed to take a small whale as it surfaced. This hunting by aboriginal peoples, which continues legally to this day, probably had no effect on any species' long-term survival. Historians agree that commercial whaling—which, in sharp contrast, has threatened virtually every species of large cetacean—did not begin until the Basques began hunting whales around the turn of the second millennium.

The Basques, who have lived in parts of what are now Spain and France for thousands of years, may be the oldest culture in Europe. Their language, unrelated to any other on the planet, is famously incomprehensible to outsiders, though their word for whale—*balea*—seems familiar enough. The Basques kept no records of their whaling activities, but we know that more than a thousand years ago they looked forward to the arrival of right whales in the Bay of Biscay each October. When the whales were spotted by a lookout on the shore, groups of men would launch their small open boats and row out toward their quarry. A harpooner stood at the bow, directing the crew to bring the boat alongside the whale. If their timing was good, the approach coincided with the animal's breathing at the surface and the harpooner could plunge the point into the whale's back. The harpoon was not designed to kill outright; rather, it was attached by rope to a large, hollow gourd. This created enough drag that the fleeing, frightened whale would soon exhaust itself as it pulled the bulky object through the water. When the whale was tired out, the men finished it off by driving one or more lances into its flesh, probing with the blade for the heart or lungs. They watched the blowholes for the surest sign that the whale was mortally wounded: blood in its exhalations. The whale was then towed to shore, where it was cut up and its blubber cooked in open pots to extract the oil.

While other Europeans may have hunted whales on a small scale in the Middle Ages, only the Basques turned whaling into an industry. They established a trade in whale oil with the people of Denmark, Britain, France and the Netherlands, who used it mainly to light their lamps. The right whale also has unusually large strips of baleen, and this flexible material could be fashioned into an array of objects, from hairbrushes and umbrella ribs to corset stays and fishing rods. There was even an important market for its flesh, which was excepted from the Catholic Church's ban on meat during feast days.

The Basque monopoly continued until the early 17th century, when a new species of whale was discovered in the Arctic. British explorers had arrived home with tales of a northern ocean brimming with what they would come to call the Greenland whale, a close relative of the right whale that is known today as the bowhead. The blubber of a bowhead is even thicker than that of a right whale, and its oil proved so valuable that the British and Dutch immediately wanted a piece of the action. In 1611, a British expedition—employing Basque harpooners, who were the only ones with the necessary skills—killed the first 13 bowheads off Svalbard, a mountainous archipelago in the Arctic Ocean, just 700 miles from the North Pole. Over the next century, Britain and the Netherlands would battle for control of these unimaginably hostile but whale-rich waters. Across the Atlantic, meanwhile, the colonists of New England began hunting right whales in the 1640s, and the island of Nantucket emerged as the centre of a new American industry. Whaling had become a multinational affair.

Both right whales and bowheads have two unfortunate characteristics that made them attractive to early whalers. They are slow swimmers, relatively easy to catch up with in a human-powered boat, and their vast oil reserves cause them to float when dead, which enabled whalers to tow their catch to shore. For these reasons, the two species sustained the Atlantic and Arctic whaling industry for more than 700 years. Humpbacks, grey whales (which today exist only in the Pacific) and others were killed when the opportunity presented itself, but before the 18th century no other whale species was targeted to anything approaching the same degree. The Basques kept no catch records so it is impossible to know how many whales they took, but by the early 1700s the animal had all but disappeared from the Bay of Biscay. Today, perhaps 400 remain in the North Atlantic. As for their cousins, the bowheads, they were wiped out in parts of their former range but are recovering elsewhere and probably now exceed 20,000 throughout the Arctic.

European whalers did not hang up their harpoons after they effectively exterminated the right whale; by 1712, Nantucketers had discovered the sperm whale, which would become the bread and butter of the industry, and famous in seafarers' lore as a ferocious monster that could reduce a ship to splinters. (Two such incidents in the 1800s, the sinking of the whaleships *Essex* and *Ann Alexander*, helped popularize Herman Melville's *Moby-Dick*, in which the *Pequod* meets a similar fate.) The sperm whale's enormous head contains a waxlike substance called spermaceti—the name was ascribed by whalers who, apparently dulled by months at sea, believed it to be the whale's semen. This unique substance, commonly called sperm oil, would become one of the most highly prized products in the industry. Originally used for making clean-burning candles and lighting fuel, it was used as a lubricant in watches and automobile transmissions, as a treatment to soften leather, and even to make soap and detergent until the 1970s.

While the distinctive attributes of right whales, bowheads and sperm whales made them the targets of whalers before the mid-1800s, the physical traits of the rorquals kept them out of harm's way. Blue whales, which were well known in the North Atlantic, can be twice as long as right whales and contain vastly more oil. (Just how much depended, of course, on the animal's size and the efficiency of the factories. A single blue whale generally produced about 70 to 80 barrels of oil, or about 3,000 to 3,300 gallons. The record holder was a single specimen caught off southern Africa that yielded 305 barrels.) But blues are much too swift to pursue in an oar-driven boat—when fleeing they can probably exceed 20 knots—and even if a whaleboat were able to catch up to one and kill it with harpoon and lance, blue whales sink when they die, and hauling a hundred-ton carcass off the ocean floor by hand is beyond the ability of even the most industrious whaler. Thanks to this lucky combination of swiftness and negative buoyancy, the largest of whales was safe until someone could devise a method for getting around these two obstacles. The solutions arrived together around the 1860s, and they would quickly bring an end to the blue whale's immunity.

Svend Føyn was born in 1809 in the Norwegian town of Tønsberg, the son of a master mariner and shipowner. When Svend was 4 years old, his father was lost at sea, and his mother found herself struggling to raise the family on her own. Føyn would always remember these years when money was tight, and from an early age he was consumed with the idea of making a fortune, which he began to do in his mid-20s. Following his father's vocation, he made a series of voyages in the 1830s to the waters between Greenland and Svalbard to hunt seals and walruses. He soon saw the potential for big profits in this enterprise, and in 1845 he commissioned Norway's first ship designed specifically for sealing, the business that would make him a wealthy man. On a sealing voyage four years later, Føyn killed his first whale—a bowhead, a species that was already rare after two and a half centuries of hunting. Føyn could see that the Arctic seas were teeming with other large whales, however. He looked with particular longing at the blue whale—a single one was worth as much as 300 or 400 seals—but the speedy rorquals were still out of the reach of hunters. More than any other man, Føyn would close the gap.

During this same period, changes in the world economy began to threaten the whaling industry. Beginning in the 1830s, new products, including kerosene, mineral oils and vegetable oils, were poised to make whale oil obsolete. After petroleum was discovered in the late 1850s, the price of whale oil fell steadily again. With stocks of the most commonly hunted whales already dwindling in many areas, making them less efficient to hunt, it looked like whaling itself might become extinct. Ironically, however, the same technological advances that rendered whale oil less important also led to new inventions that finally made it possible to kill the blue whales that had so tempted Svend Føyn two decades earlier. If whalers could tap this seemingly limitless resource, the industry would not only survive, but would grow larger than ever.

The first important step in that direction was the invention of steam-powered ships, which appeared in British whaling fleets in 1857. Six years later, Føyn oversaw the building of the first steam-driven vessel designed specifically for whaling. Almost 95 feet in length, it had a 20-horsepower engine that could do 7 knots. That's a far cry from a racing blue whale—it would take diesel engines many decades later to match that velocity—but at the time, whalers knew it wasn't necessary to pursue rorquals at top speed. If they could move stealthily, with a quiet-running engine that would not frighten the whale, they would be able to get within the range of a harpoon.

Of course, no hand-held weapon could kill a blue whale. By the 1850s, whalers realized that the future of their industry would depend on a new weapon: a harpoon that could carry an explosive powerful enough to kill large whales, and a gun that could reliably discharge it from a safe distance. American whalers, after all, knew what happened when they plunged a hand harpoon into the back of a bull sperm whale—the wounded animal could tow the whale-boat for miles, a harrowing experience they called the Nantucket sleigh ride. Better to devise a weapon that would kill, or at least mortally wound, the animal with a single shot.

The concept of the harpoon cannon had been around as early as the 1730s, but no one had yet built one that whalers would actually use. The earliest designs were awkward and dangerous, and most fired smallish harpoons that weren't sufficient to take down a blue whale. As one author puts it, "Sticking four-foot harpoons into the largest and most powerful animals that have ever lived was more suited to Gulliver's travels than modern commerce."[1] It would be at least a century before the idea showed real promise. In the 1820s, the English rocket pioneer William Congreve tested a harpoon propelled with black powder and fitted with a shell that exploded on impact with the whale. It accomplished this goal efficiently. Unfortunately, the dead whales promptly sank to the bottom. More experiments followed, and by the 1850s, an American company developed the first "bomb-lance," which was used with some success on grey whales by Charles Scammon, a 19th-century whaling captain who later re-created himself as a naturalist. In 1858, Scammon writes, a whaling brig in the North Pacific came upon a large number of blue whales and opened fire on them with the new weapon:

Ten were "bombed" by the best shooters, who affirmed that they "chose their chance," but as soon as the gun was discharged the whale would disappear, and that was the last trace seen of it, except a patch of foam, sometimes mixed with blood . . . The swiftness of the Sulphurbottom under water, as demonstrated at this time, appeared to make pursuit impracticable. Doubtless, several of those fired at received mortal wounds, or were killed outright, but their propensity to sink, and also to "run under water," baffled the skill of the whalers to secure them.[2]

As the 1860s opened, another American attempted to build a harpoon gun that would solve the rorqual problem once and for all. Thomas Welcome Roys, a captain who had three decades of experience in Arctic whaling, was something of an eccentric; he claimed, for example, that he had once ridden on the back of a whale after being dumped overboard in the North Pacific. With the help of a New York fireworks manufacturer, Roys designed a rocket harpoon that was held on the shoulder like a bazooka. From a range of up to 100 feet, the weapon fired a barbed harpoon fitted with a shell that exploded a few seconds after burying itself in the whale's body. Roys's biographer states that the old captain fired the weapon into hundreds of blue whales during live-fire exercises off Iceland. By 1866, one of his expeditions killed 49 whales, most of which were blues, but the method was so costly and labour intensive that he lost money. And, again, many of the dead whales sank before the whaleship could secure them. Roys's designs were ultimately unsuccessful, but he remained convinced that the future of whaling depended on a weapon that could kill the large rorquals. He made considerable sacrifices to the cause: before abandoning his experiments for good, Roys managed to blow off his left hand with one of his prototypes.

The harpoon gun needed a new champion, and once again it was Svend Føyn who took up the challenge. He made countless small experiments, rigorously testing new ideas for packing the gunpowder, for attaching the barbed head to the steel shaft, for setting the grenade's fuse (he used a glass vial of sulphuric acid that shattered on impact) and for securing a line that wouldn't snap under the tremendous pull of a large whale, eventually solving all of these problems. His most important contribution may have been getting the weapon off the shoulder: he invented a swivel mount that could be secured to the bow of the whale catcher, allowing the gunner to aim far more easily. His design was essentially perfected by the early 1870s, and the bow-mounted harpoon cannon remains a fixture on modern whaling vessels, improved and honed, but fundamentally unchanged. It would be difficult to overestimate the role this erstwhile sealer played in the near extermination of the blue whale. As Richard Ellis writes, "To the present day, virtually every whale killed by the hand of man has been killed by Foyn's inventions."[3]

Killing the whales was all well and good, but there was still the problem of how to retrieve them before they sank to the bottom with their precious oil. Steam power not only drove the main engines of whaling vessels of this period, but also allowed the invention of powerful winches. A line attached to the harpoon could be wound about this mighty tool, and the operator could haul it in or let it out as necessary to compensate for the movements of a whale once it had been harpooned. With a blue whale attached to one end, the strain on the line was obviously tremendous, and a sudden jerk could snap it instantly. In 1866, Roys solved this problem with his invention of a system that used a number of elastic rubber strops to lessen any sudden pull, similar to the way a fishing rod absorbs the strain on the line as it bends. If the whale died and sank, the winch and a series of pulleys could also be used to bring it back to the surface. Within two decades, whalers had figured out how to keep the whales from sinking in the first place: they pulled alongside the animal as soon as it was dead, inserted a tube into its body cavity and injected it with compressed air to make it buoyant. The last technological hurdle to hunting blue whales had been surmounted.

In 1873, Føyn obtained a patent on his entire whale-catching apparatus, seizing a monopoly in Norway for 10 years. The waters off Finnmark, the nation's northernmost county, proved bountiful indeed, and the blue whales that had once escaped his harpoons became its most prized targets. During the decade of his monopoly, the price of whale oil rebounded as the product again found a market; Føyn became a multimillionaire and the Norwegians secured their position as the world's premier whaling nation, a distinction they would not surrender until a century later, when the Japanese took it from them. Føyn's annual catch was about 80 to 100 whales, and as other companies challenged his monopoly—by this point, even some of Føyn's countrymen were beginning to think him greedy—hundreds of whales were killed each year using the new methods. Føyn's logbooks reveal that in June, July and August blues were the only species his fleets hunted, and it was the beginning of an assault on a species that would last a century.

Once Føyn's monopoly expired in 1883, it was open season on the blue whales off Finnmark, and the annual take increased dramatically as other companies clamoured for a slice of the pie. With the help of steam-powered ships and Svend Føyn's gun, the Norwegians effectively emptied the eastern North Atlantic of large rorquals in about 30 years. More than one historian has referred to that dubious achievement as killing the goose that laid the golden eggs. One solution might have been to find an egg substitute, or at least a sustainable way of collecting the golden ones; instead, the Norwegians looked for another goose. Their search, which proved to be more rewarding than any whaler could have dreamed, took them all the way to the bottom of the world.

If the blue whale is eventually to disappear from the earth, it would be safe to say that the seeds of its extinction were sown on a lonely Antarctic island called South Georgia.

Jutting out from the sea at 54°S latitude, South Georgia lies below and to the east of the Falkland Islands, more than 1,300 miles from the tip of South America. About 105 by 25 miles across at its widest, the island is a crescent studded with 11 mountain peaks that exceed 6,500 feet. While it was probably visited in 1675 by the English explorer Anthony de la Roche—he was blown off course and may have sheltered in one of its bays—the first humans known to have landed on the island were the crew of HMS *Resolution* under the command of Captain James Cook. While searching for the Antarctic continent, Cook and his men came upon the outcropping and thought at first they had reached their goal. Soon realizing they had come upon an island, they mapped its circumference, then went ashore on January 17, 1775, fired a few ceremonial muskets to claim the treeless outpost for the British and christened it after King George III. The island did not impress Cook, who described it as "savage and horrible"[4] and "not worth the discovery."[5]

While not part of Antarctica proper, South Georgia is within the Antarctic convergence, a zone that encircles the globe's southernmost land mass and marks the boundary of a distinct climatic region. Like the treeline in the northern hemisphere, the Antarctic convergence is not an arbitrary line on a map, but a natural boundary that varies between 48°S and 61°S latitude as it meanders around the continent. Along this fog-shrouded ribbon of ocean, the colder air and water from the south converge with warmer fronts from the north—mariners who have passed through it say that the drop in temperature is immediately noticeable. Within these icy waters, where the Atlantic, Indian and Pacific Oceans merge to become the Southern Ocean, tiny plankton flourish, providing food for enormous populations of krill, which once supported hundreds of thousands of baleen whales.

Because of its unique climate, South Georgia and its inshore waters are teeming with wildlife, particularly seabirds, seals and, at least originally, whales. One of the scientists with Cook's expedition remarked that if whalers in the Arctic Ocean were ever to run out of raw material, the Antarctic would supply as much as they could handle. The technology that would make Antarctic whaling possible was a hundred years away, and the comment was scoffed at. By the end of the 19th century, however, whalers in Europe were whispering stories of a vast, unexploited population of Antarctic whales. With northern stocks drying up, a new source would need to be found if the industry were to survive. But the Antarctic? Whaling in the frozen North was difficult, dangerous and costly enough, even with Arctic waters lying a relatively short sail from Norway and Britain. The Antarctic was on the other side of the world and had even more severe conditions—no one would manage to set foot on the continent until 1895. The idea of sending whaling fleets there seemed ridiculous, however many animals might be ripe for the slaughter. Nevertheless, in September 1892, a young captain named Carl Anton Larsen set out from Sandefjord, bound for the Southern Ocean. Larsen's intended quarry was not the blue and fin whales that were disappearing off Finnmark. Instead, he went in search of the right whales that many believed awaited in the frigid Antarctic seas.

Larsen was born in Tjølling, Norway, in 1860, and at the age of 32 his beard was already stiff with the salt of the Arctic Ocean. He had hunted both baleen whales and seals in northern waters when he took command of a pioneering expedition to scout for new whaling grounds in the Antarctic. Larsen's ship, *Jason*, passed South Georgia in November and headed south to the waters east of the Antarctic Peninsula, which protrudes from the continent like an icy finger. He looked for right whales in vain, but he killed a number of seals to make the trip worthwhile, and he watched countless blue whales ("as many as twenty at a time"[6]), fin whales and humpbacks swarm around his ship. Ill equipped to hunt them, he could only look with longing, as Svend Føyn had done on his own early sealing voyages.

Larsen returned to the Antarctic the following season, but again he left the rorquals undisturbed and concentrated on seals. He continued to be struck by the number of large whales he encountered, however, and became convinced that these animals were the future of the industry, an argument that would be strengthened by the fact that four expeditions to the south between 1892 and 1895 resulted in the killing of exactly one right whale. But if Larsen was a visionary, many of his backers were more conservative. Although blue whales had been killed off Finnmark and Iceland for decades, whaling company officials in Britain and Norway continued to focus their efforts on the moribund right whale.

By the dawn of the new century, polar exploration was all the rage, and beginning in 1901 Larsen made a series of voyages with the newly launched Swedish Antarctic Expedition. During one of these, he dropped a group of scientists on an island in the Weddell Sea, where they planned to spend the southern summer. On his return voyage, Larsen visited South Georgia, which he saw as an ideal location for a whaling station. During his earlier, unprofitable journeys, Larsen had recognized that Antarctic whaling could only succeed if it had a local base of operations—why not this island, with sheltered harbours, shores that were ice-free all year, land that was suitable for buildings and plenty of fresh water? With that idea filed away, Larsen returned to pick up the Swedish scientists several months later, but his ship became trapped by ice and eventually sank, forcing the men to spend the winter of 1903 in a makeshift stone hut on a barren island. After a terrible ordeal during which one of the crew died, Larsen and the others were rescued by an Argentine vessel and taken to Buenos Aires, where a feast was held in his honour.

Never one to dwell on past misfortune, Larsen immediately began looking for Argentine investors to help him establish a whaling station on South Georgia: "I ask youse ven I am here vy don't youse take dese vales at your doors—dems vary big vales and I seen dem in houndreds and tousends."[7] A group of businessmen soon established the Compañía Argentia de Pesca, and Larsen returned to Norway to look for backers in his own country. He met with little success, but he did find experienced Norwegian crews willing to man the company's three vessels, which set out on the first commercial whaling expedition to the Antarctic in 1904. Their operation was based out of a South Georgia harbour called Grytviken, meaning "Cauldron Bay," a reference to the detritus left there by sealers a century earlier.

The first whales to be targeted by the new industry were not the giant blues (though 11 were killed in that first season), nor the right whales that had enticed the whalers in the first place. Rather, it was the slow-moving humpbacks that first died in great numbers. Pesca, as the company became known, took just 147 in its debut season, but six years later, with British and Norwegian companies now stampeding to the ice and establishing other shore stations on South Georgia, whalers killed an astounding 6,197 humpbacks. Once that species had been decimated, blue whales were next in line at the slaughterhouse door: the 1914–15 season saw more than 2,300 winched onto the decks at South Georgia, and another 1,800 killed in the nearby South Shetlands. During these early years, the bounty was so great that there was no incentive to painstakingly extract every drop of oil. Only the thickest blubber was removed, with the rest being left to rot, a practice that sickened even the most unemotional whaler and eventually led to regulations requiring that the companies make use of the entire carcass. "There are few instances in mankind's long history of greedy exploitation," writes Richard Ellis, "that demonstrate such wanton, senseless destruction of a natural resource."[8]

With a population of about 150 men, South Georgia had become the most southerly inhabited land on the planet. Larsen introduced reindeer to the island in 1911 to provide fresh meat, and the animals continue to thrive there, but there were few other reminders of the outside world. Even for hardy Norwegian and British seamen, life on the island could be lonely, gruelling and dangerous. The profits were enormous, though, and while the individual whalemen did not reap the same returns as the companies' owners and investors, they made a comfortable paycheque, which was easily saved because there was no place for a man to spend his money. As long as whaling was expanding, there was a constant need for skilled labourers to go to the Antarctic, and the work was considerably more lucrative than comparable landlubbing trades. The whaling industry, which had seemed close to collapse not a decade earlier, had never seen anything like the boom that followed the opening of the Antarctic. The only question was how long it would last.

Basque whalers made their profits by selling meat for food and oil for lighting and dressing wool. The British and Dutch who pursued the bowhead whales of the Arctic also made fortunes from baleen, which was extraordinarily versatile in the age before plastic. The American whalers of the 19th century sold countless barrels of sperm oil to be turned into candles and used as a lubricant and tanning agent. These whale products filled legitimate needs in the marketplace that could not be met through other means. But by the first decade of the 20th century, there was no significant world market for the meat or baleen of rorquals, and synthetic products, fossil fuels, domestic animal fats and vegetable oils had become far more important than whale oil for all of its traditional uses. What, then, drove the economic engine that led to the unimaginable slaughter of blue whales in the Antarctic? The answer is that hundreds of thousands of these animals died so they could be turned into margarine and soap.

There was great demand for both of these products by the end of the 1870s, especially after butter substitutes had become a staple in European diets. But for decades whale oil was unsuitable for making hard soap or margarine, because it is a liquid at room temperature. (About one-third of whale blubber is oleic acid, the same unsaturated fatty acid found in olive oil.) Then, in the early 1900s, a seemingly innocuous discovery had enormous consequences for the blue whale. A German chemist discovered a process called hydrogenation, which can convert oleic acid into a saturated fat, and suddenly it was possible to use whale oil as raw material for margarine and hard soap. The product became attractive again, especially when producers of domestic animal fats and vegetable oils could not keep up with the demand of soap makers and food manufacturers. The recently discovered Antarctic whaling grounds, brimming with blue whales that could easily supply 15 or 20 tons of oil each, looked irresistible, and between the 1909–10 and 1913–14 seasons, the production of whale oil almost tripled to meet the voracious demands of the market.

The arrival of the war in 1914 slowed the whaling industry as ships and men were needed for the military, but the harvest continued, with an average of almost 3,900 blue whales killed during each of the four wartime seasons. During the austerity of these years, Europeans became more willing to hold their noses and consume whale products, and companies such as Lever Brothers (later to become Unilever) bought up as much oil as they could; in 1916, consumption of margarine in Britain surpassed that of butter for the first time. In addition, the war created an unlikely new market for whale oil. During the manufacturing process, the glycerine that forms the chemical backbone of whale oil is separated from the fatty acids that are the active ingredient in soap. When glycerine is combined with two nitrogen atoms and an oxygen atom it makes nitroglycerine, a highly explosive compound that became a mainstay of the armaments industry during the war. It was a sad irony that as humans were blasting blue whales with grenade harpoons in the Antarctic, a by-product of those same animals was blowing up soldiers on the battlefields of Europe. As Richard Ellis put it, the whales "gave their lives so that humans might die."[9]

After the war, margarine tubs rather than soap dishes emerged as the final resting place for most of the Antarctic's blue whales. By the mid-1920s, improvements in hydrogenation made whale-oil–based margarine more palatable and economical; by 1930, chemists had devised a method for removing the fishy taste and smell completely, so margarine could be made using whale oil almost exclusively. Soap manufacturers, by contrast, preferred to use other fats, such as palm oil. Public prejudice against whale oil continued, and correspondence from Lever Brothers' chairman in 1926 reveals that the company was doing what it could to distance itself from whaling, even though the two industries were intimately related: "Is it not counter to Lever Brothers' policy to approve any wide dissemination of the suggestion that whale oil and Lever soaps and food products have an association?"[10] Lever Brothers' concern was not that consumers might be appalled that Antarctic whaling threatened to wipe out several species of magnificent mammals in order to make mundane household products. The problem was simply that in the 1920s Britons still thought the stuff was smelly and unhygienic. This denial would continue for decades, and most people outside the whaling, margarine and soap industries had little or no idea precisely why the species was being hunted to near extinction.

Within a few seasons of Captain Larsen's opening of Grytviken on South Georgia, early Antarctic whalers began experimenting with a new way of processing whales in "floating factories." For centuries, the boilers and separators used to process whales had been built at shore stations, but by the early 1900s, companies had begun moving their operations onto large vessels, usually converted oil tankers or cargo ships, so they could be moved from site to site as needed. In the beginning, these floating factories were some 1,500 gross tons and required a crew of about 60. They were typically moored in a harbour; the whales were towed alongside them to be flensed—that is, have their blubber stripped away—and the cut-up fat and meat was lifted on board by derricks and winches and dropped into the cookers. As whale stocks around South Georgia and other Antarctic islands began to dwindle, whalers realized that their future depended on cutting the tether that kept them to inshore waters. By the 1920s, whaling companies had developed the first true pelagic factory ships—vessels capable of operating in the open ocean with a fleet of catcher boats—and by the time they had matured, these enormous ships could displace more than 13,000 tons and carry a 500-man crew. For the blue whales that roamed hundreds of miles offshore, away from the earliest Antarctic expeditions, the modern factory vessel was what one historian called "the technological development that spelled final disaster."[11]

After returning to Norway from South Georgia in 1914, Carl Anton Larsen had repeatedly brought up the idea of leading the first pelagic whaling expedition, but it was not until 1923 that he had everything in place. Accompanied by five catcher boats, Larsen skippered the factory ship *Sir James Clark Ross*, which, at 12,000 tons, was the largest whaling vessel of its day, and headed south via Tasmania into the Ross Sea. Larsen had hoped to find new stocks of humpbacks, but when the fleet made its first kill in December, the quarry was a blue whale. A young Australian journalist, A.J. Villiers, accompanied the expedition and later recorded the scene in bombastic style:

> Quietly and surely [the gunner] took long and careful aim at the interminable gray flank slowly turning before him. At last his finger twitched ever so slightly, and with a boom and a roar, a deluge of flying, twisted pads, a reek of explosive, the great shell-pointed steel harpoon flew out, and in a flurry of boiling foam the stricken leviathan sounded deep into the depths, in a terrible effort to rid itself of the burning steel. But his doom was sealed.[12]

The catcher boat soon killed three more blue whales, "all bulls, between eighty-five and ninety feet long,"[13] and towed them back to the *Sir James Clark Ross* to be flensed. But there were problems. The factory ship had been built to work in a harbour like its predecessors, and the men had been accustomed to cutting up whales that lay alongside a moored vessel. It proved immensely difficult to do this in a rolling sea, particularly in hand-numbing cold. The other hurdle was that the lines and winches were designed to handle humpbacks, and they simply weren't able to deal with a 100-ton blue whale. "It was a hopeless task," Villiers writes, "and it had to be abandoned." The crews soon realized they would have no choice but to tow their whales to a sheltered spot, anchor the *Sir James Clark Ross* and process the whales much like they had on earlier floating factories. Larsen recognized that if blue whales were going to be hunted in the open ocean, someone would need to come up with a way of getting past these limitations. The incentive was certainly there: one of the blue whales killed on this pioneering voyage was reportedly 106 feet long. It must have made an awful lot of margarine.

The solutions that would ultimately enable humans to pursue blue whales in the open ocean were not long in coming, though Larsen himself did not live to see them. The captain died in his cabin from angina on December 8, 1924, and his body was taken home to Sandefjord, where he received the largest funeral ever held in that town. The year after Larsen's death, spurred by the potential that his voyage had showed, the Norwegians launched the *Lancing*, the first factory vessel with a stern slipway: a hatch in the back of the ship led to a ramp, allowing the crew to drag a whale right onto the deck, where it could be safely and easily cut up. (This seemingly obvious idea had been conceived of some decades before, but earlier prototypes interfered with the ship's propellers and rudder.) On its voyage to the Antarctic, the *Lancing* spent July to September in the Atlantic waters off Congo, testing its new equipment on humpbacks, and almost 300 were hauled through the slipway with great success. When the ship eventually killed its first blue whale, however, it was only with enormous effort that the crew managed to winch it aboard. Eventually, stern slipways were studded with semicircular ridges and wet down to reduce the friction. The final difficulty the *Lancing* experienced—attaching a cable to a whale's tail in a heaving sea—was soon overcome by the invention of the whale claw, a clever device designed to fit over the tail flukes and tighten automatically when tugged.

For as long as blue whales had shared the earth with humans, they had been afforded some protection by their speed, their tendency to sink when dead, their enormous bulk and their habit of living hundreds of miles offshore. By the end of the 1920s, human ingenuity had found ways of overcoming all of these natural defences. Now it was left to human greed to look after the rest.

———

Hans Bechmann still has a clipping from the published diary of a naturalist who visited the Antarctic in 1912–13. It was faxed to him in 1989 by a friend who recognized a reference to Bechmann's father:

The stern slipway and the whale claw finally ended the blue whale's immunity.

"Here at South Georgia we have had a tragedy in a fatal accident to Captain Beckman [*sic*], who had the reputation of being the best whaling gunner in the world." In November 1912, Johan Ludvig Bechmann, a pioneer of Antarctic whaling who "was idolized by the members of his crew," had been at the bow of a new steam-powered catcher, *Don Ernesto*, when the recoil of the gun caused the mount to break. The captain fell onto the deck below and suffered a fatal skull fracture. "He was only 27 years old," the entry continues, "and leaves a wife and four children."[14]

Lethal accidents aboard whaling vessels were not unusual during this time. As many as 200 men died during the first 18 seasons in the Antarctic, many of them finding rest in South Georgia's permafrost, while others were buried at sea. Johan Bechmann, however, was no common whaleman, and it was ordered that his body be embalmed and carried home to Norway to receive a hero's funeral. The order came straight from Captain Carl Anton Larsen himself.

The youngest of the four children who lost their father that day would grow up dreaming of having his own whale catcher. "Not only my father, but two of my uncles were also gunners," recalls Hans Bechmann, now in his mid-90s and perhaps the oldest surviving whale gunner who worked in the Antarctic. "It was a big whaling family. I was hoping I would be a gunner too, from the time I was 15 years old." Bechmann would go on to spend 27 seasons in the bow, eventually becoming one of the most sought-after gunners of the mid-20th century. His career aligns almost perfectly with the era of blue whale hunting in the Antarctic. When Bechmann first went to sea in 1927 at the age of 15, pelagic whaling was in its infancy and the Southern Ocean still had hundreds of thousands of baleen whales. At the pinnacle of his career as a gunner, he was taking 10 or 12 whales on a good day, and he killed more than 300 in his best season. But by the time he loaded his last harpoon in 1965, Bechmann could see the end had come. "In the last season," he remembers, "we didn't see any blue whales."

After a dozen years working various jobs aboard whaling vessels, Bechmann got his first opportunity as a gunner in 1937. "The first year I was down there we didn't have any speed—we could only do about 10 knots, so it could be difficult if the whales started running. By the last season we could do 16 or 17 knots in the catchers." Despite the relative slowness of the catcher boats in the 1930s, fleets in the Antarctic killed 14,826 blue whales in Bechmann's

The swivel-mounted harpoon cannon, invented by Svend Føyn, has been a fixture in the bow of whale catchers from the 1870s to the present day.

debut season, a testament to the skill of the gunners, who achieved godlike status in their hometowns. As far back as the Yankee whaling era, it was customary for whaling crews to receive a base wage, plus a bonus for each whale killed. Both figures, of course, varied according to the man's position on the ship, and no man was more highly paid than the gunner. In the early 20th century, a seaman on a Norwegian catcher boat made about 45 kroner per month, with a bonus of 5 kroner for every blue whale killed, and 3 kroner for other species. The gunner, who was also typically the captain of his vessel, earned 120 kroner per month, plus another 50 for each blue whale, 40 for a fin whale, and so on, which amounted to five or six times more than a seaman's salary over a full season. During a successful career a whale gunner could amass a considerable fortune.

During his early years in the Antarctic, Bechmann was employed by Christian Salvesen and Company, a firm with a long connection to whaling. Its founder came to Scotland from Norway in 1846 and later opened a business that imported whale oil from Svend Føyn. In the 1890s, Salvesen got involved in its own whaling activities and became the first British whaling firm to enter the Antarctic; it would remain one of the world's premier whaling enterprises until the 1960s. Though based out of the Scottish port of Leith, Salvesen maintained a close association with its North Sea neighbours, and while Antarctic whaling crews were often a mix of both nationalities, the gunners were almost exclusively Norwegians.

There was a reason that gunners were held in such high esteem: killing a blue whale with a harpoon required immense skill. The hunt would begin when the man in the barrel—the catcher boat's equivalent of a crow's nest—would shout down to the crew after spotting the blow of a whale. As captain, the gunner would pilot the ship to within range of the animal before handing control to a helmsman and scrambling along a duckboard bridge to the gun platform. Bechmann explains that he would try to get within 165 to 300 feet, with the catcher travelling slowly to avoid alarming the whale. Once he was within range, he took aim and pulled the trigger. Accompanied by a tremendous bang and the smell of gunpowder, almost 6 feet of steel would fly from the gun, trailed by a heavy line called a forerunner. Typically the gunner aimed at the middle of the whale's body, Bechmann says, where he believed the heart was. About five seconds after the harpoon struck its target, the grenade exploded with a dull thud, and the whale shuddered as its vital organs were blown apart.

Bechmann says he never lost a blue whale once he had it in range, and one shot from his gun was usually enough to kill it outright—two reasons why he was among the greatest in his profession. In the hands of men with lesser skills, the harpooning of a whale could be a drawn-out, gruesome affair. On his voyage with the *Sir James Clark Ross*, Villiers describes what could happen when a shot was less than perfect:

> The blue whale is a tremendous creation—the largest in the world—and it takes a lot to kill one. If it is not struck in a vital spot it will require two, and even three, great steel harpoons

and bombs to finish it, despite the terrible destruction of its organs by the bursting of the soft-iron bombs. Sometimes the harpoon glances off the curved back and the bomb explodes harmlessly in the air, or entering the blubber a rib is struck, and the harpoon is bent double, drawing out ere long. If the head is struck, the harpoon does not enter at all, and if the shot has gone too near the tail, the whale will tow the steamer furiously about for hours.[15]

The death throes of a blue whale could be prodigious. The record holder is an animal harpooned off Newfoundland in 1903. When first struck, it towed the catcher at 6 knots for most of the day, even with the ship's engine churning at half-speed in reverse. The crew tried to fix the line at the stern so they could run the engines forward at full throttle, but when they managed to do so, the whale almost pulled the ship under the water. The tug-of-war continued all night, before the animal finally became exhausted and the men launched a small boat to finish it off. The whale's agony lasted 28 hours.

In the lore of 19th-century American whaling, tales abound of ships being rammed and sunk by sperm whales after they had been wounded, or in some cases even before they were harpooned. The exploding grenade certainly made non-lethal strikes less common, but it was not unusual for a wounded whale to turn on the catcher boat—though whether the animal targeted the ship or simply struck it unintentionally in its thrashing is impossible to know. J. N. Tønnessen and A. O. Johnsen, in their definitive *History of Modern Whaling*, report at least five occasions in which a vessel was actually sunk, although they do not name the species involved. Whalers seem to agree that blue whales, despite their superior bulk, were easier to kill than sperm whales, which will not surprise any angler who knows that a smallmouth bass can put up a greater fight than a walleye twice its size. Certainly blue whales would have been capable of sinking a small catcher, but whether any actually did so is unlikely. Nonetheless, in his account of the *Sir James Clark Ross*'s Antarctic expedition, the intrepid Villiers writes:

> The blue whale, wounded to the death, has been known to charge madly at his puny assailant and send the little vessel, a broken mass of scrap-iron and matchwood, reeling to the depths. Many are the names of whalers, spoken of with dread in the quiet of the night, which have left the mothership in excellent weather never to return. Strange are the stories of the finding, weeks afterward, of the bodies of great whales bearing the unmistakable brand of a marked harpoon from a lost whaler.[16]

This little gem no doubt entertained the readers who devoured the dispatches Villiers sent home to Australia, but it is nonsense. Either he couldn't resist sprinkling his book with a few tall tales, or else the whalers he travelled with put one over on the credulous scribe. For his part, Hans Bechmann never saw or heard of such an incident during a career that spanned five decades.

Indeed, if blue whales had been able to fight back they may not have died in such astounding numbers. With the major technological advances of modern whaling now perfected and the skill of the gunners finely honed, the 1930s were an era of unmatched slaughter, and the blue whale cemented its role as the ultimate quarry in the animal kingdom. In the 1929–30 season, whalers killed more than 19,000 blues, as many as all other species combined. The following season saw the record catch of any single species: 29,409 blues, representing 80 percent of the worldwide harvest of whales. Perhaps another 10 percent were struck and killed, but not recovered, meaning that the death toll was more like 32,000. To put that number in perspective, it is probably two or three times the global population of blue whales today. During the eight seasons beginning in 1932–33, whalers in the Southern Ocean averaged more than 15,000 blue whale kills a year.

The assault on the Antarctic blue whale may have been the greatest war humans have ever waged against an animal. North

Americans killed hundreds of millions of passenger pigeons in the 18th and 19th centuries, wiping out what was once the most common bird on the continent, but the animal's extinction also came about through deforestation and disease. Even the dodo—an icon of extinct creatures—was less likely to have been killed by humans than by the alien species they brought to Mauritius. Hunters with rifles have wiped out many large mammals, of course, from the zebra-like quagga in southern Africa to the tigers of Bali and Java, but these were localized populations that made relatively easy prey. Blue whales, in sharp contrast, are so large and elusive that it took whalers a thousand years just to develop the technology needed to kill one, and the systematic hunting of the species required sending vessels and crews to the most remote place on earth. Perhaps most remarkable, the real carnage occurred during an extremely brief and bloody period—the golden age began in 1928 and lasted a mere dozen seasons, during which more than 191,000 Antarctic blue whales died. It would take another terrible war, this one launched by Nazi Germany in 1939, to bring about a ceasefire.

THREE

BACK TO THE ICE

O N THE MORNING of September 20, 1940, the factory ship *New Sevilla* set out from Liverpool, bound for South Georgia. During the previous two days, trucks had carried load after load of beef and mutton aboard to sustain the whaling crews during their months in the ice, and the ship was loaded down past the legal limit. The huge vessel, weighing more than 13,800 gross tons, had been built in 1900 by White Star Line, the company that owned the *Titanic*, and had originally taken passengers from England to Australia. In 1930, she was converted to a factory ship and began making regular trips to the Antarctic. During the decade that followed, the *New Sevilla* was part of a whaling fleet that killed about 149,000 blue whales in the Southern Ocean.

With the Battle of the Atlantic now well underway, whaling vessels had to travel in convoys on their way south, and the *New Sevilla* was accompanied by 47 other ships. Around 8 p.m., some 10 hours after leaving Liverpool, a tanker in the convoy was struck by a torpedo from a German submarine and blown to pieces. Minutes later, the factory ship itself was targeted by *U-138* and took a hit below the funnel. Alex Peterson, a British whaler aboard the ship, later remembered the moment: "The alarm bells went and everyone had to get on their lifebelts and get to hell up on deck. When the submarine put the second one into us, it was right into the engine room and that just finished her completely."[1] Peterson was thrown into the water, but managed to hang on to a capsized lifeboat; eventually he grabbed the cargo net that hung alongside one of the convoy's other ships. All but two of the *New Sevilla*'s crew were saved before the ship finally sank.

Peterson and the other survivors would eventually get to South Georgia aboard another vessel, but the 1940–41 whaling season was severely hampered by the war. Two of the main whaling nations during the previous decade—Germany and Britain—were completely engrossed in the conflict, while Norway, the other major player, was occupied by the Nazis after April 1940. Most of the world's whaling vessels were needed for military purposes—catcher boats were commissioned by navies, and factory ships that once supported the enormous carcasses of blue whales were instead used to transport supplies. The Japanese reportedly even used the stern slipways of their factory ships to launch and retrieve miniature submarines. The floating factories, however, were so large and slow that they became easy pickings for U-boats. In October 1942 alone, three were torpedoed and sunk in the North Atlantic with the loss of 161 men. In all, of the 37 factory ships that had been involved in whaling before the war, just seven remained when it was over. Germany, Britain, Japan and Norway also lost more than a third of the 312 whale catchers that had operated during the last peacetime season.

It is a tragic irony that it took a war involving nations from six continents to halt the slaughter that was taking place off the seventh. Fewer than 5,000 blue whales were killed in the Antarctic in 1940–41, mostly by the Japanese, who had not yet entered the conflict, and the catch was less than 200 during the next two seasons combined. From 1943 to 1945, Norwegian expeditions killed about 1,400 blues, but the collective catch during the five wartime seasons was a bit more than half what it had been in 1939–40 alone. Not since the first decade of the century had so few whales been hunted, and many in the industry hoped the reprieve might give blue whale stocks a chance to recover from the carnage of the 1930s. It was not to be.

While whale oil had been extremely valuable during the First World War, it played a much smaller role in the Second. In the postwar years, however, raw material from baleen whales—both oil and meat—became economically important once again. In the UK and western Europe, the demand for imported fat was particularly high after the lean years of the war, and to stimulate production the British government fixed the price of whale oil at £75 per ton, almost double what it had been in 1939. Whale oil produced by Norwegian companies was fetching £120 per ton by 1950. At this time, whaling companies had to know these lofty prices could not continue—vegetable oils were already threatening to overtake whale oil as the preferred raw material for margarine—and only the wilfully blind believed the stocks of Antarctic whales could endure anything like the slaughter of the previous decade. But the economic incentive was so high that the whalers went back to the ice immediately after the war ended.

Germany and the United States, both of whom had joined the Antarctic whale hunt in the late 1930s, never went back. Norway, the UK and Japan, however, rebuilt their fleets and were joined for the first time by the Netherlands and the Soviet Union. South African companies, which were also working in the Antarctic before the war, continued whaling there until the late 1950s. While the number of blue whales killed in the Southern Ocean would never again reach pre-war levels—strictly because of scarcity, not for lack of effort—it would come close, with almost 9,000 taken in 1946–47. By this time, fin whales were the most commonly killed species, but blues were still the preferred catch, since they yielded far more oil than a fin whale of the same length. In both 1949–50 and 1950–51, while fewer than 7,000 blue whales were killed in each season, the species still accounted for one-third of the total mass of all whales taken in the Antarctic—more than half a million tons annually. A gunner would happily turn his harpoon away from a pair of fin whales to take a single blue.

James Meiklejohn spent 10 seasons in the Antatctic, beginning in 1951. By the time he left, blue whales had all but disappeared from the southern Ocean.

During the summer of 1951, the factory ship *Southern Venturer* was docked outside Newcastle in northern England. Operated by Christian Salvesen and Company, the *Southern Venturer* measured 540 feet from stem to stern, and was about 75 feet wide. The sight—and the stench—of the enormous vessel still hasn't faded from the memory of James Meiklejohn, who joined its crew at the age of 17, having never before been to sea. "Just coming onto the factory when she was lying in docks in South Shields, it still smelled from the last season—even though they had hosed it down with caustic, the smell was all over the town." Meiklejohn was born in Scotland in 1934, and was planning to continue his education when the summer was over. "I was fond of languages—we had learned Latin and French at school—and I thought I would do something with that." But then a recruiter for Salvesen suggested he sign on for a year of whaling. "I said yes, hoping it would be only one season, and my summer holidays ended on the *Southern Venturer*. I sailed with them over to Norway, then three weeks to Aruba, and then down to the ice." What was supposed to be one season turned into 10—five on board the *Southern Venturer* and five on South Georgia, where he overwintered four times.

That first season, the teenaged Meiklejohn was mainly responsible for working on the boilers and cleaning the evaporators, but because of his education he was often called up to assist the chief engineer. "I calculated the fuel used by each whale catcher, and how many nautical miles the *Venturer* sailed by the slip percentage—how many times the propeller went round." At that time, most of the crew belowdecks were British, but the officers were all Norwegian, and they took kindly to Meiklejohn's interest in their language. Soon after arriving in the Southern Ocean, Meiklejohn saw the men take a 50-foot sperm whale, the first he'd ever seen. Weeks later, the sight of his first blue whale would remind him of the sea creatures he had read about in Jules Verne's stories—it was measured at 102 feet. "They had to cut off the fins to get it up the slipway. Can you imagine that? It looked like a three- or four-storey building."

While the major advances in factory ships and whale catchers came before the war, the whaling fleets of the post-war era had become even more efficient. In what was a typical arrangement, the *Southern Venturer* was teamed with 13 catchers and 2 buoy boats, usually older catchers whose job was to retrieve the carcasses and tow them to the factory ship, which freed up the gunners to pursue other whales. Once an animal was harpooned, the crew of the catcher would inflate the body cavity to keep it afloat, mark it to indicate which catcher made the kill and then attach a lamp and radio buoy to make the whale easier for the buoy boat to find. Usually the crew cut the tail flukes off, since the natural undulations of the sea could cause a dead whale to "swim" along the surface, making retrieval more difficult.

Once the catch arrived at the factory vessel, the decks came alive with the sounds and smells of a whale being worked up. The crew began by attaching a claw to the whale's tail and using a steam winch to haul it up the slipway onto the aft deck, where the flensers were waiting. Using rounded knives on the ends of poles—the tools look like field hockey sticks—the flensers stripped the blubber from the back of the whale. Generally, one man would begin on each side of the mouth, while a third stood atop the whale and walked along its spine. Being careful not to cut into the meat—the blubber of

Even the most jaded whaler could not help but be impressed by the huge size of blue and fin whales as they lay on the flensing deck of a factory ship.

a blue whale ranges from 2 to more than 12 inches thick—each flenser made incisions along the length of the whale's body. The men then attached hooks to the strips at the head end, and a winch behind the tail would tear the strips of blubber off the whale, as though they were peeling a 90-foot banana. As the back blubber separated from the meat beneath, it made a sound that one whaler described as "a crackling like a bonfire of pea-sticks."[2]

The next task was to turn the whale over so the flensers could remove the belly blubber. They accomplished this Herculean feat with a pair of winches: they placed one wire across the top of the body and attached it to a pectoral fin, while a second wire, wound about a winch on the opposite side of the ship, went underneath the whale and was secured to the other fin. With both winches working together the whale could be slowly rotated, and when it was belly up the flensers went to work again, rolling back the thinner layer of blubber on the ventral side. A blue whale's blubber makes up about a quarter of its overall body weight, so on a good-sized animal these strips of blubber could easily weigh 20 tons, yet six flensers could complete the task in only an hour. In this short time, whatever majesty a blue whale might have retained on the deck of a factory ship disappeared as the blubber, now cut into manageable chunks, was fed into the cookers. The biggest challenge came from the animal's tongue. "It was a huge mass of slippery, jelly-like matter which took up a deck space of about sixty-four square feet," writes the Salvesen whaling captain W.R.D. McLaughlin in *Call to the South*.[3] Alex Peterson, the whaler who was aboard the *New Sevilla* when it was torpedoed, remembers that wrestling a blue whale's tongue into the boiler could have mortal consequences. "Once you got it started you had to run like hell to keep clear of it. If you got caught with a bit of it you could say goodbye."[4]

With the flensing complete, the crew hauled the whale through "Hell's Gates," the narrow division between the aftdeck and the foredeck, and the lemmers took over. A lemmer was a butcher of whales; his job was to dissect the stripped carcass so the body parts could be processed. Using hooked knives and giant steam-powered saws, the lemmers removed the meat and separated the bones and connective tissue. "There was great activity there," Meiklejohn remembers, "and when the whaling was good it looked like Dante's *Inferno*." The high-quality meat was often frozen, while the remainder went into the boilers and pressure cookers to extract the oil. It was a remarkable feat of human cooperation and technological efficiency that a blue whale could be rendered in about seven hours.

While oil remained the most important raw material extracted from whales, meat also became important in the post-war years, although only in Japan—campaigns by food producers to encourage Europeans to eat whale meat all failed. People in Europe found the idea unpalatable, not the meat itself. Most people who have eaten the flesh of baleen whales report that it tastes at least as good as beef, perhaps better, although the blue whale's diet of krill can leave a fishy odour that some find distasteful. During their months in the ice, many whalers embraced the chance to eat the meat of their catch. W.R.D. McLaughlin shares a story about a conversation that took place on his ship:

"What the hell's in these bangers?" inquired one whaleman, as he poked his head into the shop one day and saw the butcher making sausages.

"Just the usual," replied the butcher. "Whale-meat and pork."

The whaleman looked around the immaculate shop and saw large heaps of whale-meat and pork, chopped up and ready to go through the mincer. "Meat and pork," he exclaimed. "I suppose you put in about fifty per cent of each?"

The butcher grunted. "You're quite right," he said. "Fifty per cent of each. One blue whale to one pig!"[5]

In the UK, however, where people regularly eat blood pudding, pork hocks and kidney pie, the meat of a whale was deemed unappetizing. This was perhaps because earlier generations of British whalers had told stories of the rotting, stinking carcasses they had processed at shore stations. Modern technology now enabled whale meat to be kept fresh, of course, but the prejudice seemed to linger. This refusal to eat whale meat was surely not because of a moral aversion—widespread public awareness about the plight of endangered whales would not come for two decades, and in any case, Europeans could not have rejected blue whale meat out of concern for the species while at the same time spreading the animal's fat on their bread.

Indeed, while Britons did not eat whale meat themselves, they happily fed it to their pets, though most likely had no idea of the source of their tinned cat and dog food. Around 1970, when the researcher George Small, while writing his landmark book *The Blue Whale*, contacted a Norwegian company that sold whale meat to a British pet food manufacturer, the firm refused to disclose the price it received; by this time it would have had some cause to worry about a public outcry. Small later learned from a whistle-blowing employee that the company collected £75 to £100 per ton, which it considered an excellent rate. "It did not admit, naturally," he wrote, "that profits were being maintained at the risk of biological extinction of the blue whale that was being killed to feed cats and dogs in England."[6]

Norwegian whaling companies, like their British counterparts, also sold some of the meat from their catch to the Japanese fleet before leaving the Southern Ocean, but in relatively small quantities. They brought almost none of the meat home. While some people in Norway enjoy whale meat even today, the market in that country has never been large. In Japan, however, the product was enormously popular, and remains so today. In the 1950s and 1960s, the Japanese produced more whale meat than beef, and it sold for about a third of the price. Their access to a large and lucrative market for whale meat was a major reason they were able to make profits from whaling well after many European companies had given up. No one knows how much of the meat was derived from blue whales—Japanese whaling companies, well aware that they are pariahs in much of the world, have been less than forthcoming—but given the high price they received for the meat, the incentive to target the largest species is obvious.

There were other less important resources extracted from blue whales. The residue left after the oil was extracted from the meat could be dried to make a high-protein animal feed. Before vitamin A could be synthesized, the whale's liver was sometimes processed to extract this nutrient. Once the oil was removed from the blubber and any meat cut and prepared, the remainder of the carcass was often processed to make guano and bone meal, which were sold as fertilizer. Somewhat ironically, given that it was once one of the most valuable of whale products, the baleen of blue whales was almost always discarded. Despite having a mouth large enough to comfortably fit a cube van, the blue whale produces smaller and inferior baleen to that of bowheads or right whales. In any case, by the mid-20th century, the market for this substance was nil.

———

Even after South Georgia's shore stations were rendered obsolete in the age of factory ships, the island remained a crucial part of whaling operations in the Southern Ocean. Salvesen's activities were based in Leith Harbour on the northern coast of the island, and James Meiklejohn decided in 1953 that he would overwinter there after the *Southern Venturer* steamed for home. It would mean 22 consecutive months away from family and friends, but the pay was good and the taxes low. During the winter, the catcher boats were brought into dry dock to be painted and maintained, and the small shore crew did a variety of tasks to prepare for the next season. "The winters in South Georgia were fine—I enjoyed them," Meiklejohn says. "It wasn't fifty below like it was on the Antarctic continent. But sometimes we had three or four metres of snow there, so the whale-catcher people sometimes did nothing other than shovel snow off the decks of the ships, or the catchers would have gone under water. We had a Kino [a cinema] and we went there every Wednesday, Saturday and Sunday. There was always something to do: people built models, or made scrimshaws. I took classes in Norwegian on Wednesdays, and if there was no rum ration, that usually went well."

Before 1958, no mail made it to South Georgia until the first factories arrived for the start of the whaling season, but that winter a boat arrived for the first time with bags of letters. As the station's technical secretary, Meiklejohn was responsible for divvying up the mail and handing it out to the lonely men, who would gather in the main mess in anticipation. Toward the end, there would be a small group of men around him who had received no word from home. "I'd notice that they'd had no letters, and I could see the looks on their faces, and I'd say to myself, 'For God's sake, I hope they get a letter.' Then we'd get the last bag, then the last bundle with the elastic band around it, then the last letter, and there'd be maybe three people there. And one would say, 'Christ, I knew she was unfaithful.' Many times I'd take them up to my room and give them a drink and say, 'Cut out that nonsense. There's probably another bag.' You can imagine living with six men in a room, and five get 30 or 40 letters, and one gets nothing. There were tragedies there." During those long winters on South Georgia, depressed whalers were known to hang themselves or take their lives in other ways—

Meiklejohn knew of men who simply disappeared and were never found. Not all of the Antarctic's casualties were whales.

In May 1958, while James Meiklejohn was living on South Georgia, Gibbie Fraser wrote to Salvesen to ask for a job on a whale catcher. Later that year, though not quite 16 years old, the boy who had never before left his home in Scotland's Shetland Islands would be on a ship bound for the Southern Ocean. Though he couldn't have known it at the time, Fraser would be witness to the final years of blue whale hunting in the Antarctic.

"I had always wanted to go on a whale catcher," Fraser recalls from the same house in West Burrafirth where he wrote that letter half a century ago. "A man I knew was very enthusiastic about the whale catchers, and he described them in great detail. He would draw pictures of them for me and explain how everything worked. When you spoke to some of the other men who were whalers, especially some of the guys who worked in the factory ships, they'd say, 'Oh, no, you're much better on a factory ship, don't go on those whale catchers.' But I wanted to try it."

Throughout the whaling era, some of the hardiest seamen of them all came from Shetland. At a latitude of 60°N, the islands are the most northerly outpost of the United Kingdom, less than 150 miles from the Norwegian coast. Shetland was a Norwegian colony from the 9th century until 1472, and during the Second World War, the British aided the resistance in occupied Norway with a series of exceedingly dangerous missions across the North Sea in fishing boats, an operation nicknamed the Shetland Bus. In the post-war years, then, Shetlanders and Norwegians were natural allies on the whaling grounds of the Antarctic.

Nowadays, Fraser can hop in his car or onto his motorbike and drive to the capital of Lerwick to run an errand, but in the 1950s that was impossible. Few people in Shetland owned a car, and buses and ferries connecting the islands' major centres ran perhaps once a week, if they ran at all. Most of the roads were unpaved until the late 1960s or early 1970s, when an oil boom brought Shetland out of its economic slump. "Whaling was a very popular job at the time because it suited the islands so well," Fraser says. "We would leave in September or early October and we arrived back again in early May, so it took us away during the harsh winter months. We were home to help with the sheep and the crop work, and any fishing was done in the summertime." It says something about Shetland's climate that whalers went to Antarctica to escape their own winter.

"I had no relatives in the whaling," Fraser goes on, "but I knew a lot of people who went, and every year some of the boys would leave school and the next thing you'd heard they had gone to the Antarctic. I must say I felt a bit jealous from the age of 12 and on. So when I was 15, I wrote to Salvesen just after the whalers came home in May, and I got a letter back saying that they would consider me for employment." That August, he received another letter saying he was to present himself at a hospital in Lerwick on a certain date for a chest X-ray. So he and a friend who had also applied made their way to the capital and arrived at the hospital, letters in hand. "We found ourselves among a group of big, burly guys who we didn't know. Some of them looked a bit elderly, but of course they were no more than about 45 or 50."

After their medical exam, Fraser and his friend met some older boys who invited them to a Lerwick pub called The Lounge. "We said we were kind of young for that, and they said, 'Oh, never mind, nobody's going to say anything.' So we went along and had a few pints of export ale and thought we were really hardened whalers." At the pub, Fraser had his first encounter with John Thompson, a legendary whaler from the island of Yell who was nicknamed "the Arab" for reasons no one remembers. Thompson, a huge bearded man with thick glasses and a hearty laugh, was about two-thirds of the way through his pint when he suddenly made a rumbling noise and then casually turned away from the bar and threw up all over the floor. "He drew his sleeve across his mouth to dry things off, and then downed the rest of the pint," Fraser remembers. "Then he pushed

Gibbie Fraser (top left) and his fellow mess boys on their first trip to the ice.

the glass across the counter to the barmaid and said, 'Same again.' She put her hand to her mouth and slid out the back. I just turned to my friend and said, 'God, what have we let ourselves in for?'"

Fraser was eventually given a job as deck mess boy on the 200-foot *Southern Broom*, a whale catcher that had once been a corvette in the Royal Navy. Then known as HMS *Starwort*, she had done escort duty in the Mediterranean and in 1942 had depth-charged a German submarine. Years later, Fraser tracked down a photograph of the surrendering crew of *U-660* waiting to abandon ship. The corvette had been refitted with a new foredeck and a flared bow, which now sported a new kind of weapon: a harpoon gun. Crewed by 19 men—4 British and 15 Norwegian—the *Southern Broom* had a steam engine powered by two sets of boilers and could do 18 knots at full throttle—fast enough to catch up to the blue whales of the Antarctic.

Fraser's job was to set the tables and serve the food, to keep the mess room and galley clean, and to wash down the cabins, alleyways, radio room and storerooms, but there was plenty of opportunity to learn other jobs. "You could do anything aboard a whale catcher apart from shoot a whale. So of course, being young and keen and eager, I volunteered as much as I could. The spirit on a whale catcher was that you were part of a team. Once you harpooned a whale, the quicker you got everything done, the better. You wanted to get away after another whale, because that would up your bonus. So they would send you down to the hold to fetch a grenade, detonators or various pieces they might need. We were always dashing about whenever there was anything that needed to be done."

During Fraser's first season, 1958–59, most of the catch were fin whales—the *Southern Broom* found and killed just "one lonely blue whale," he remembers, and the other 11 ships that made up the fleet did not kill any. (Other operations in the Antarctic managed to kill 1,076 blue whales that season, down from 1,660 the previous year.) "But the last year I was on that catcher," he says, "there was one day when there were blue whales all over the sea. In fact, the top whale catcher had 16 or 17 for the day. We had around 12 or so. The last one we got that day measured 96 feet—it was a massive animal. One harpoon and it just turned over dead. We actually shot across the back of another whale to get this one because it was just so much broader than the rest. We towed it in to the factory ship ourselves— I think the gunner wanted to see it out of the water. They pulled it up onto the deck and when the flensers climbed up on top of that thing they looked like little spiders—it looked nonsensical."

The crew still did their long-range spotting of whales from the barrel, some 65 feet above the deck. "There was an intercom system on board, with a speaker in the barrel, another at the gun platform, one on the bridge and one in the engine room. If you were in the barrel and you saw whales off the starboard bow, or port bow, or whatever, then you'd press the button and relay that to the engine room and they would be ready to increase revolutions. You only steamed at about 12 knots when you were searching for whales, but when you saw whales and were going for them, they pulled out all the stops." While early catchers had to pursue blue whales stealthily because they could not match the speed of an animal fleeing for its life, modern vessels were capable of 17 or 18 knots. Now the hunting technique changed: catcher boats tried to get the whales to "run," knowing they would eventually become exhausted, after which the crew would move in for the kill.

While the good gunners rarely missed a whale once they were close enough to fire the harpoon, there were still many that evaded the catcher boats. "I think some of them were quite clever, really," Fraser says. "They would come up for a few breaths, and just when you were getting within range, they would go down and not come up again until maybe 10 or 20 minutes later, and then they'd be astern of the ship. Many a day was spent trying to spook them into running. Others would run right away—you'd come up among a group and get the first one, and then they would set off. That was the duty of the man in the barrel: to keep his eye on the direction the rest of the whales took. I remember one day we chased a whale in a straight line for four hours. It left us for a little while, but of course, it was a warm-blooded animal and a steam engine isn't. We gradually overtook it, but it was a long time before we could see that we were catching this whale. Everyone was saying there would be no barrels of oil left in that one."

Since the heyday of blue whale hunting in the 1930s, the techniques used to shoot the animals had not changed significantly. Hans Bechmann, who began his career as a gunner before the war, remembers that the harpoon cannon was essentially the same when he retired in 1965. Alf Mathisen, a distant relative of the great Bechmann, joined the catcher boat *Southern Soldier* as a gunner in 1955, and says it was still the same cold, wet job it had always been. The major difference now was that the whales were more scarce, and during the perpetual daylight of the Antarctic summer the catcher boats ran all day and night in pursuit of those that remained. But while other tasks aboard the ship could be shared, there was only one gunner. "There was no one to relieve you," Mathisen says. "Sometimes you were out there for 24 hours, or even 30 hours. I remember one day we started chasing at two o'clock in the morning, and I went until eight o'clock that evening without eating. I ate quite a lot for supper that night." During his best season, 1958–59, Mathisen shot almost 200 whales, mostly fin and sei, but also a small number of blues. He retired from whaling at the end of the following season after watching his bonus dwindle along with the stocks of whales.

While the harpoon gun would not change significantly for the rest of the whaling era, companies did experiment with new technologies to make killing more efficient. However skilful a gunner might be, and despite his genuine desire to minimize the animal's suffering, the agony of a harpooned whale could be disturbing to witness. One company, Hector Whaling, reportedly spent over £100,000 on experiments to come up with a way of humanely killing whales by electrocution, something that had been tinkered with since the 1930s. By the mid-1950s, some people in the industry still believed the technique had potential, and the guns of at least two catcher boats were fitted with harpoons that did not explode on impact. Instead, a jolt of electricity travelled through the line into the whale's body. The electrical harpoon ultimately failed, however. The designers were never able to adequately integrate an electrical wire with the harpoon line, and in any case it was difficult to determine just how much current would do the job. Too little and the whale didn't die; too much and its valuable flesh was cooked.

As it became more difficult to find large whales in the Southern Ocean, Salvesen and other whaling companies also tried using helicopters and airplanes to spot the animals from the air. The *Southern Venturer*, in fact, was one of a small number of factory ships with a helipad at the stern. The experiments never worked as well as the companies had hoped—poor weather often limited their usefulness. "Another problem was the identification of the whales," Gibbie Fraser says. "The people flying the helicopters and doing the spotting didn't really know what kind of whale they were seeing. They would say there's a pod of 20 or 30 whales at such-and-such a position, and two or three catchers would go there and find that they were humpbacks, which were out of season. So that wasted a lot of time and fuel."

At least one technological advance did significantly change the way whalers did their job after the war: sonar. While primitive sonar devices were built to detect submarines as early as 1915, the technology was much refined during the Second World War, and when Antarctic whaling returned to pre-war levels in 1945–46 the new catcher boats were equipped with distinctive metal domes under their hulls. When used to hunt blues and other large baleen whales, sonar—which was known to the British as asdic, an acronym for the Anti-Submarine Detection Investigation Committee that developed it—did not work like a modern fish finder, scanning the waters to locate a potential catch. Rather, the catcher boats used the device to track whales they had already sighted from the barrel; the pinging also apparently spooked the whales into coming to the surface, and often made them run. Before sonar, a good part of the gunner's skill involved predicting where a whale would surface, and in directing his ship to that position. This new technology allowed the whalers to follow the whale underwater, eliminating much of the guesswork.

The flipside was that a poor asdic operator could severely handicap the gunner by sending him on a series of wild goose chases. Alf Mathisen learned this cruel fact during his first season. "It was very difficult to be a new gunner so late in the whaling," he says, "because the new gunners always got the oldest boats. The good gunners also got the good asdic operators, while the one I had on board couldn't find any whales at all." (Once, he remembers, his operator was a former mess boy who had spent just four days learning his craft.) Although no one at the time understood how baleen whales hear, some whalers believed that sonar caused the animal considerable suffering. As one wrote in 1962: "When asdic is used by hunting catchers it is not surprising that the whales react violently, even at considerable distances. Perhaps the whale experiences the pain in the head which we get from an almost unbearably loud noise."[7]

New technologies would ultimately have little impact on blue whales, which by the end of the 1950s were so rare that only 1,200 or so were killed each year, and the hunt focused on fin whales. Predictably, these became scarce too. During the 1964–65 season, a Norwegian company hired Hans Bechmann to scout the waters below South Georgia for any sign of whales. Aboard the whale catcher *Thoris*, the veteran gunner sailed down to South Orkney, then west along the ice edge to South Shetland and on to Cape Horn before returning to South Georgia. During the 12-day journey,

Bechmann says, he saw the blow of a single fin whale. By the time the *Thoris* returned to port, the British and the Dutch were out of the Antarctic whaling business, leaving only the Norwegians, Soviets and Japanese to fight over the sei and minke whales, once considered too small to be worth the effort.

It is easy to understand why whalers returned to the ice immediately following the war: the high price of oil made whaling an attractive business, and although stocks were declining, there were still enough large rorquals in the Antarctic to make the enterprise profitable. What is harder to figure is why whaling companies continued to pursue blue whales well into the 1960s, despite the obvious fact that the species was headed toward oblivion. By that time, blues made up such a small portion of the overall catch that a ban on hunting them may well have been inconsequential to the whalers' profits. But the companies did not see things that way. Throughout the post-war period, whaling companies insisted that they had to continue their activities or they would face economic ruin. They pointed to a declining price for whale oil after the mid-1950s, as well as increasing labour costs and lower productivity due to the shrinking populations of whales. Given these threats—which were no doubt real—the companies argued that they could not agree to a ban on hunting the largest of all whales, and they resisted every attempt to protect the species. "But were those claims justified?" asked author George Small. "Was the blue whale slaughtered almost to the last individual for reasons of economic necessity? Or, was it for little more than profit that could safely be labelled economic necessity because no one bothered to question the claim?"[8] After doggedly tracking down the statistics, Small concluded that the companies still would have been profitable if they had concentrated on less threatened species. "The companies as a whole would not have faced anything approaching economic hardship, let alone disaster, if they had ceased killing blue whales after 1956. Their claims were false . . . Why then did the blue whale have to die?"[9]

Small's question may have been rhetorical, but there is an answer, albeit a perverse one. In coldly economic terms, it actually did make sense for whalers to kill as many blue whales as they could in the short term rather than maintaining a smaller sustainable hunt. Blues are a long-lived species with a low reproductive rate, and the maximum sustainable harvest never would have been more than 3 or 4 percent—about 4,000 to 5,000 whales annually. That yield would have kept the stocks healthy, but also would have required sending enormously expensive fleets to the Antarctic indefinitely. If all the companies cared about was maximum profit, it was wiser to take as many as they could in as few seasons as possible, thereby keeping costs to a minimum, and then to invest their revenue to provide an ongoing income. Think of whales as cash and consider this: if someone offered you $1 million now or $40,000 annually for eternity, which would you take? As long as you average at least 4 percent a year on your investments, taking the lump sum is the better deal. The whaling companies understood this, and they acted accordingly.

Salvesen sold the *Southern Venturer* to the Japanese in 1961. Two years later, James Meiklejohn, who had once worked in its engine room, moved to Tønsberg, a town of some 32,000 people and the birthplace of Svend Føyn himself. Meiklejohn married a Norwegian woman and settled down to a life on dry land. He continued to work for Salvesen for another four years, helping to close out accounts and sell off the company's fleet of whale catchers, after which it got out of the industry for good. (Today, Christian Salvesen is a successful logistics company with more than 13,000 employees.) "I was the last of the Mohicans," Meiklejohn says, "the last one to be involved in Salvesen's whaling." He vividly remembers the day in 1967 when the former whaling ships left the naval repair base where they had been moored. "I locked the gates and handed over the key to the naval commander for the last time." In 1984, Meiklejohn established the Salvesen Ex-Whaler's Club, which included 247 men united by their experiences in the Antarctic. As of 2008, fewer than a third of those original members survive, the oldest being the 96-year-old former gunner Hans Bechmann.

As for the blue whales of the Southern Ocean, their destruction was almost complete. In all, about 330,000 were killed during 60-odd years of whaling in the ice, and when a committee was organized to estimate the remaining population, its report in 1963 was bleak: perhaps 600 were left alive in Antarctic waters.

FOUR

ABOUT-FACE

ONE OF THE GREAT DISGRACES of 20th-century whaling is that it almost wiped out nature's largest animal creation before humans had even begun to understand it. Hundreds of blue whale carcasses were dissected and studied in the early days of whaling, leading to a reasonably thorough grasp of their anatomy. But what people knew about living, breathing, diving, feeding and mating blue whales remained embarrassingly inadequate. "A friend of mine, who is one of the foremost cetologists of our time," wrote the conservationist Farley Mowat in 1972, "recently summed up the state of our knowledge in these words: 'The little we biologists know about whales in life would hardly provide enough material for an essay by a high school student.'"[1]

Beyond the occasional measurement of a carcass, no nation made any sustained effort to study the whales they were killing until the British emerged as pioneers of this new science after the First World War. During that conflict, the demand for edible oil and nitroglycerine led to the slaughter of 175,000 whales, including more than 15,000 blues, and by that time even the most short-sighted whaling companies realized they would have to do something to conserve the stocks. And they couldn't do that unless people understood more about the whales' life history, movements and ecology.

When the First World War was over, the British Colonial Office formed a committee that included several scientists who would become the first to gather reliable scientific data on the largest inhabitant of the world's oceans. In 1923, this committee acquired the RRS *Discovery*, which had carried Robert Falcon Scott and Ernest Shackleton to the Antarctic on their historic expedition in 1901. The 736-ton, three-masted vessel had also done duty as a cargo ship for the Hudson's Bay Company in Canada, and later ferried military supplies to Russia during the war. The Colonial Office repaired the ship, reinforced it for duty in the ice and equipped it with the latest oceanographic instruments, after which it steamed to the Antarctic. The proud vessel soon attached its name to the group itself, and the newly christened Discovery Committee began its ambitious effort to study whales and their habitat.

The Discovery Committee was hardly a conservation group in the modern sense—the British authorities were devoting their resources to preserving the whaling industry, not the whales themselves—though it would be unfair to dismiss the undertaking as cynical. The Apollo program may have been driven by national chauvinism in the 1960s, but that does not mean its scientists were uninterested in exploring the moon for its own sake. In the same way, the Discovery Committee's researchers missed few opportunities to indulge their curiosity about the fascinating and mysterious animals they were studying for the first time.

The committee's work began in early 1925 when it set up a Marine Biological Station at Grytviken harbour on South Georgia, opposite the grave where Ernest Shackleton's body had been interred three years earlier. Almost all whaling in the Antarctic to this point was still done at shore stations, supplying the scientists with more specimens than they could handle. After recording the species of each whale hauled onto the flensing platform, they noted its sex and carefully measured it—not only its overall length, but many other dimensions, including the tip of the rostrum to the blowholes, the width of the tail flukes, and the eye to the ear. They counted and measured the baleen plates, tallied the number of ventral grooves under the jaw, and even counted the number of hairs on the chin. They cut open the whales' stomachs to see what they had eaten; in the case of blue whales, this was almost always a single species of krill, *Euphausia superba*. The scientists also examined the whales' ovaries or testes and removed fetuses from pregnant females. The work, in the words of one of the scientists, was "both laborious and odorous,"[2] but it provided the data needed to understand the blue whale's gestation, growth rate, and breeding cycle, "since the amount of destruction which can be carried out without consequences serious to the industry is clearly dependent on the rate of replacement."[3] The data also helped determine the proportion of old to young animals being killed. If the percentage of juvenile whales was increasing, it would indicate that the stocks were being depleted. By April 1927, the Marine Biological Station had examined about 1,700 carcasses, and when it finally closed four years later the scientists departed with data on some 3,700 whales.

While its work was focused in the Antarctic, the Discovery Committee also sent researchers to two whaling stations in South Africa: one at Saldanha Bay, on the Atlantic coast, and the other at Durban, on the Indian Ocean. Both stations were regularly taking blue whales, usually in the winter, leading to suspicions that southern Africa might be a destination for blues migrating up from the Southern Ocean. Two of the scientists who spent time in both South Africa and the Antarctic were Neil Mackintosh and John Wheeler, who published a massive paper on blue and fin whales in the first volume of the *Discovery Reports*, which began appearing in 1929. This paper contains so much useful data on the external and internal characteristics of the large rorquals that today it is still considered the classic reference for blue whales.

Meanwhile, the RRS *Discovery* was busy surveying the waters around South Georgia and Britain's other Antarctic territories, where it was soon joined by two other vessels. *Discovery II*, specially constructed for the committee, was a steel-hulled, oil-burning steamer that could venture farther and more safely into the ice than its predecessor. From 1931 to 1933, with the help of regular refuelling from whaling vessels, *Discovery II* circumnavigated the Antarctic continent, the first time such a feat had been accomplished in winter. Once or twice each day, the crew lowered bottles into the sea to obtain samples from various depths, sometimes as much as 3 miles. By measuring the salinity, temperature, acidity and oxygen content of the water, the scientists began to understand the horizontal layers and currents of the Southern Ocean. It was their work, in fact, that led to the discovery of the Antarctic convergence. They also collected samples of plankton, and these gave them a deeper appreciation of how this resource is not evenly distributed in the sea—which is why whales are not evenly distributed either. While this was not technically difficult, crew members had to do it with bare hands while being battered with freezing spray in a region where gale-force winds were routine.

The Discovery Committee's work also involved an experiment designed to learn more about the movements of Antarctic baleen whales. To explore this perennial mystery, the committee engaged its third vessel, the *William Scoresby*, in a program of full-time whale marking in 1934. This involved bringing the ship within 70 to 80 yards of a surfacing whale and, with the help of a shotgun, firing a steel marker into the animal's blubber. Each of these marks was stamped with a serial number and a message: "Reward paid for return to Discovery Committee, Colonial Office, London." The idea was that when a marked whale was eventually harpooned, the flensers or lemmers who carved the blubber and meat would find the mark and send it back to the scientists along with details about when and where the animal was killed. Sometimes, however, the steel tag turned up in a cooker, or even in a piece of stored meat, which cast doubt on which whale it had actually come from. (Whalers were paid £1 for each mark returned, and the committee allowed them to keep it as a memento after they logged the data.)

Marking whales proved to be difficult. One of the early designs looked like a fearsome thumbtack, with a point that penetrated just a couple of inches and a disk that remained outside the body. These quickly fell off, and not one was ever found in a dead whale. Other prototypes corroded in the whale's body, rendering the serial number illegible. Eventually, the scientists settled on a 10-inch stainless-steel tube that completely buried itself in the blubber and penetrated the muscle. This worked brilliantly, and by 1939 the *William Scoresby* and a small fleet of hired catchers marked more than 5,000 whales, from which they ultimately recovered almost 300 marks. Blue whales were harder to pursue than fin whales and humpbacks, and the total number successfully marked was only 668, of which 38 were eventually retrieved. (Researchers made more whale-marking expeditions between 1953 and 1967, successfully tagging another 1,500 blues and recovering about 50.) Almost half of the recovered marks were found during the same season—one on the same day—though eight were returned a year later, and some were at large for much longer. At least five blue whales carried the piece of steel around for more than a decade; the most elusive was marked on March 8, 1935, and killed almost 13 years later on January 20, 1948.

The inherent limitation of the Discovery Committee's marking program was that the whales were both marked and hunted in the southern hemisphere summer, so this method wasn't likely to shed light on the animals' presumed winter migrations. Had a whale marked off South Georgia in January been killed in July off South Africa, that would have proved at least some Antarctic blues migrated north. But that never happened. Whales were regularly killed over a thousand miles from where they had been marked, but the line between those points never spanned more than 8 degrees of latitude. The farthest north any pre-1939 Discovery mark was ever recaptured was 54°S.

The recovered marks did demonstrate some trends in the movement of Antarctic blue whales, however. Those marked near South Georgia, for example, were routinely found closer to the pack ice later that same season, suggesting that they were not lingering near the island, but simply passing it during their southward summer migration. Two blues were marked off South Georgia in one season and killed the following year in almost the same spot, indicating that at least some individuals return to the same feeding sites in different years. Finally, the enormous longitudinal span between some of the marks and recaptures—one whale moved halfway around the globe, from about 81°W to 88°E—confirmed that blue whales roamed vast ranges in the Southern Ocean.

The Discovery Committee continued its work for more than a quarter century before it was wrapped up in 1954. It produced 37 volumes of valuable research, the last of which was not published until 1980. Many of the methods it pioneered in studying whales—from sighting surveys, to marking, to exploring the ocean conditions that give rise to krill swarms—remain fundamentally unchanged by later generations of biologists. In these respects, the Discovery Committee was an unqualified success. In the end, however, the committee fell well short of its ultimate goal, which was to preserve the stocks of blue whales and other rorquals in order to allow the whaling industry to continue sustainably. There were at least two reasons for this failure. The first was a scientific error: despite examining thousands of carcasses, many of which were pregnant females, the committee's scientists concluded that blue whales reach sexual maturity at age two. Because of this belief, whalers remained absurdly optimistic about the species' ability to recover from their harvest. It was not until much later that scientists understood that sexual maturity comes no earlier than age five, and probably around ten. By the time they figured this out, blue whale populations had been reduced to a fraction of their original numbers.

The committee's second tragic failure cannot be blamed on the biologists or oceanographers. It was simply that its work was ignored. In a review of the committee's findings published in 1937, one commentator wrote that "enough knowledge is now available to secure the preservation of the whale."[4] That may have been true, despite the misunderstanding about their age at sexual maturity. When Neil Mackintosh later reflected on his career, he too mused that "the Discovery Committee's work on whales has generally helped to give a solid foundation for the regulation of whaling."[5] But if the committee laid a foundation, the only thing built upon it was a slaughterhouse. During the committee's first season in 1925, whalers killed about 6,000 blue whales. Six years later, after many of its important findings about blues were published, whalers quintupled that death toll to 30,000. Blue whales continued to die in hopelessly unsustainable numbers until the Second World War, and by 1950—the year Mackintosh delivered his head-in-the-sand lecture about the Discovery Committee's supposed success—the catch figure was again just over 6,000, because the species had been decimated. If the aim was to bring about the rational regulation of the industry and preserve its prize quarry, the committee was an utter failure. It would soon be followed by a litany of others.

The first international attempt at regulating Antarctic whaling followed the two most disastrous seasons of blue whale hunting in history—for both the hunters and their quarry. Some 48,000 blues were killed in 1929–30 and 1930–31, far and away the worst one-two punch the animals ever suffered. The problem, though, was not that whaling companies worried about dwindling stocks after these banner seasons. On the contrary, the problem was that whale oil was so plentiful that the market flooded, prices plummeted and some raw material had to be discarded after it spoiled in warehouses. The industry quickly realized it would have to curtail its supply. "Probably all the factory-ships will have to lay up for a year," wrote A.J. Villiers in 1931, "not to give the whales a chance, but to give the market one."[6] Now the blue whales of the Southern Ocean were not just dying so that humans could make them into margarine and soap. Many were dying for no reason at all.

Before this period, the Norwegian and British governments had made occasional attempts to regulate whaling off their own coasts, usually to mollify local fishermen who claimed the whalers were interfering with their livelihood. In the Antarctic, however, where vessels flying several flags worked the same waters, national policies carried no weight. In 1931, the League of Nations took steps to bring together the major whaling countries with an eye to hammering out the industry's first international agreement. The document that came out of those talks, the Geneva Convention for the

Regulation of Whaling, was signed in January 1932 by 26 nations. It gave complete protection to right whales, forbade the killing of all calves or females accompanying them, required whaling vessels to reduce waste by using all parts of the whale, and obliged that crews be paid according to the amount of oil the whales yielded, rather than the number of individual animals killed. This last article was supposed to discourage gunners from shooting immature whales. But its effect was to place an even bigger target on the backs of blue whales.

The convention was the first multilateral whaling regulation to be signed—and it was the first to be ignored. The whole process was suspect from the beginning, since Norway and the United Kingdom were the only signatories that had a significant stake in Antarctic whaling at the time. The representatives from countries such as France, Italy, Turkey and Poland supplied moral support, but given their lack of whaling interests, this was meaningless. Meanwhile, several nations that would later become active in the hunt—Japan, in particular, but also the Soviet Union, Chile and Argentina—did not sign the convention. Germany signed, but did not ratify it. On top of all this, there was no way of enforcing the regulations or penalizing nations that violated them. During the three years it took to ratify the agreement, 52,000 more blue whales died in the Antarctic.

The whaling nations tried again in 1937, when they met in London, to prevent the collapse of the industry. In many ways, the agreement that followed from these talks, the International Convention for the Regulation of Whaling, was an improvement over its predecessor, although that is not saying much. It gave belated protection to grey whales in the western North Pacific (this population is still critically endangered today). It established minimum size limits, but they offered little respite for immature whales: no blue whale under 70 feet was allowed to be taken, for example, but female blues in the Antarctic do not reach sexual maturity until 77 or 78 feet. The agreement also banned factory ships from operating outside of the Antarctic, the North Pacific and the Bering Strait, the only places where significant pelagic whaling was going on. The convention also shortened the whaling season so that it began on December 8 and closed on March 7, which still allowed plenty of time for factory vessels to fill their holds with whale oil. (The Japanese, preoccupied with their war in Manchuria, were not a party to the agreement, so they were not bound by these dates anyway.) Another protocol, passed a year later, established a sanctuary in an area of the Antarctic where whales were so scarce that almost no hunting took place.

It is impossible to overstate the futility of conservation measures in the 1930s. Some of the regulations were absurd—the selective ban on pelagic whaling, for example, was like making it illegal for Americans to pick pineapples except in Hawaii. During the debates, a small number of delegates argued that only much stricter controls would prevent the elimination of the most vulnerable species. They argued for strict quotas on the number of blue whales, fin whales and humpbacks that could be killed each year. At the very least, the companies needed to limit the number of vessels they sent to the Antarctic, since their destructive power was now enough to empty the seas if it was not checked. Most people probably understood this. The problem was that each whaling country had no reason to trust the others to play by the rules. Norwegian companies knew that if they kept to the quotas, but other nations did not—and both Japan and Nazi Germany openly refused to consider any of the proposed limits—then they would lose profits and fail to protect the whales at the same time.

The Second World War halted the carnage; virtually no blue whales were taken in 1941–42 and 1942–43. However, five months before the D-Day landings, all the nations that had signed the 1937 convention, with the exception of Germany, were back at the conference table in London. The agreement they signed finally established a quota for whales in the Southern Ocean. Rather than setting separate catch limits for each species, however, the quota system was based on a dubious measure known as the blue whale unit (BWU). Whalers had come up with this unit based on a rough estimate of how much oil each rorqual yielded. One blue whale was the benchmark; it equalled 2 fin whales, 2.5 humpbacks or 6 sei whales. This scheme, set out for the first time in 1944, not only failed to protect Antarctic blue whales, it virtually guaranteed their destruction.

The system worked like this: before each season, the whaling nations agreed on a certain number of blue whale units that would represent a total quota for the industry. Once a week each expedition would radio the Bureau of Whaling Statistics in Norway and report the number of BWUs they had taken. For example, if the week's catch consisted of 12 blues, 16 fin whales and 12 sei whales, that added up to 22 BWUs. Officials would tally these results and project the date on which the quota would be reached. They would then notify the factory ships, and all of them had to stop whaling by that date.

Unfortunately, this system was fatally flawed in several ways. First, it ignored the fact that not every species was equally vulnerable. Blue whales were much closer to collapse than either fin or sei whales, but the BWU system offered no incentive to take the more plentiful species. On the contrary, given the choice between shooting one big blue or chasing down two finbacks, catcher crews would choose the former. In both the whaling and fishing industries, there is an assumption that a species will never be harvested to the last individual, because as soon as it is scarce it becomes uneconomical to pursue. The blue whale unit demonstrated why that idea breaks down if vessels are pursuing a more abundant species at the same time. "There is a chance of using one species to subsidize the extermination of another," wrote one observer in 1962. "The fin whales are far more abundant, but the gunners prefer to take blue, so that the last blue whales might be shot by an expedition relying mainly on fin."[7]

The second shortcoming of the quota system based on BWUs was that it encouraged whalers to scramble madly for the largest share. Assigning a percentage of the quota to each expedition or each country—something the whaling nations would not accept until the early 1960s—would have allowed each to hunt at a relatively leisurely pace. Instead, catcher boats and factory ships oper-ated all day and night whenever weather permitted, with the goal of taking as many whales as they could before the deadline. The increased competition forced companies to come up with newer and better ways of killing. Factory ships started carrying helicopters and airplanes to spot whales. Catchers got faster, and their hulls were fitted with sonar domes to track the animals and drive them to the surface. Buoy boats retrieved the carcasses so the top gun-ners could shoot more whales. The whaling companies poured big money into improving their fleets, which motivated them to insist the quotas remain high so they could earn a return on their invest-ments. It was a vicious circle with the whales trapped in the centre.

The quota system also rewarded unscrupulous expeditions that fudged their numbers—they might report a small blue whale as a fin whale, for example, so it would count as only half a unit. The Soviets, in particular, were accused of either underreporting their catches, or else overstating the numbers in an effort to bring a quick close to the season. After the competition left, the Russian factory ships would have the whaling grounds to themselves.

The final failing of the BWU quota system was perhaps the most predictable: the quotas were perennially too high. During the first season of the new rules, the total was set at 16,000 units. Many sci-entists knew this was already far too high to be sustainable—even half that total would have strained the stocks—though many in the industry argued for a quota of 20,000 BWU. The whaling countries paid lip service to the idea that the quota would be reviewed annu-ally and adjusted if necessary, though it was not reduced signifi-cantly until 1963–64. In the four seasons before the quota system was finally abolished in 1971, the figure was between 2,700 and 3,200, but the whalers fell a few hundred short each year. Can there be a greater act of stupidity than setting a quota so high that there were not enough whales alive to meet it?

After the Second World War, the whaling nations decided they needed a permanent regulatory body to replace the *ad hoc* negotia-tions. In 1946, they gathered in Washington, D.C., for a meeting that would lead to the formation of the International Whaling Commission (IWC). In December of that year the founding members—all of the major whaling countries except Japan, who would join in 1951—drafted the International Convention for the Regulation of Whaling, which was to "provide for the proper conservation of whale stocks and thus make possible the orderly development of the whaling industry." There was still time to prevent the collapse of the Antarctic blue whales: almost 7,000 were killed in 1946–47, suggesting the population was large enough to recover if the species was protected immediately. Even agreeing on a sustainable quota would have been a momentous improvement over every previous international agreement. Instead, the whaling industry chose to kill even more whales. The same year the convention was signed, two new countries began sending factory vessels to the Antarctic: the Netherlands and the Soviet Union. As Richard Ellis writes:

> By 1946, the framers of the International Convention for the Regulation of Whaling knew all about the destruction of the right whale and the disappearance of the Arctic bowheads. They had before them the bitter documentation of the decline of the Pacific gray whale. And yet they persisted: they designed a system to oversee the destruction of the remaining whales of the world.[8]

The IWC, created to uphold the convention, first met in London in 1949. Within five years it included a Scientific Committee made up of 11 biologists from seven member states whose role was to provide the data necessary for the rational regulation of the industry. Not surprisingly, however, the scientists were ignored and short-term economics routinely won out over long-term conservation. In 1953, they pointed out that catcher boats in the Antarctic were averaging one blue whale for every nine days' work—down from more than one whale a day in 1934. They noted that whalers were catching 15 times more fin whales than blues, a clear indication that the latter species was in steep decline. They also proposed shortening the season for hunting blues, and closing an area of the Antarctic where the whales were particularly scarce. All the member nations agreed on the first point, but the Netherlands refused to accept the closure of any part of the Southern Ocean. As a result, the proposal quickly died.

This incident, and many others that followed it, underlines the most glaring weakness of the IWC. While the British and Norwegians had argued as early as 1946 that the commission's decisions should be binding on all members, that is not how the IWC was set up. Instead, any member country can object to any decision within 90 days; once it has done so, it is not bound to abide by it. This effectively gives veto power to all the members. If every country except the Netherlands agreed to stop killing blue whales in one part of the Antarctic, the Dutch would simply—and legally—have the area to themselves. The other nations could hardly be expected to go along with this. IWC members are well aware that giving everyone veto power renders the commission almost impotent, but the alternative is worse. If every majority decision were binding, a nation that felt continually thwarted would simply leave the commission and continue its activities with no regulations at all. Chile and Peru did just that. Both signed the 1946 convention but did not ratify it, so they operated outside the IWC's control until they joined in 1979. Norway, the Netherlands and Iceland have also walked out temporarily, while Canada, which grants licences to Inuit communities to hunt bowheads, quit the IWC for good in 1982. Other countries that allow aboriginal whaling, such as Indonesia and Tonga, have never been members.

Indeed, whether the IWC can exercise any real control over its members is debatable. This isn't the fault of the secretariat, nor even the national delegates; after all, as a voluntary international organization, the IWC cannot impose meaningful sanctions on member countries. One thing the commission could have done more effectively, however, is convince its members to carry an international observer aboard all factory vessels to make sure that regulations were followed. The members had discussed this as early as

1959, but then proceeded to squabble over the details for a decade and a half before finally implementing the idea in 1972. By that time, the United States, which—along with Britain, Australia and other former whaling nations—had adopted a conservationist stance with the zeal of a reformed smoker, took its own steps to add teeth to the IWC's regulations. The American government threatened to stop importing fish from any country it deemed to be in violation of whaling restrictions, and to curtail the amount of fishing that nation could do in U.S. waters. It has since flexed these economic muscles against Japan, the Soviet Union, Chile, South Korea, Peru and Spain.

Until the U.S. took these measures, the IWC floundered every time it had an opportunity to protect the dwindling resources of the industry it was supposed to preserve. At its 1955 meeting in Moscow, a proposal to ban the killing of blue whales in the North Pacific was shot down not only by Japan and the Soviet Union, but also by Canada and the United States, where a small number of coastal whaling stations remained open. Then the members re-opened the Antarctic whale sanctuary that had been set aside in 1938. The following year, the IWC decided that blue whales were a lost cause, and any effort to conserve baleen whales in the Southern Ocean should focus on finbacks. "The year 1956 was possibly the last year when the blue whale could have been saved by giving it complete protection," George Small reflected. "If the killing continued unabated much longer the coup-de-grâce would surely come."[9]

In 1960, after a decade of ignoring its own scientific advisers, the IWC hired a trio of independent biologists from non-whaling nations. This aptly named Committee of Three was to assess the stocks of each baleen whale species and figure out a sustainable quota for each. The scientists presented their final recommendations at the 1963 meeting and announced that no more than 5,000 blue whale units, a third of the current figure, was reasonable. Norway and Britain agreed and proposed a quota of 4,000; the Japanese insisted on 10,000. It is not difficult to guess who got their way. The scientists also recommended a total ban on killing blue whales, which they believed numbered just 600 in the Southern Ocean. Most of the members were prepared to agree, but again Japan held out. Its vessels had recently discovered a region in the southern Indian Ocean that still contained a relatively healthy population (they later declared them a new subspecies, pygmy blue whales). The other whaling nations saw little alternative but to agree to leave this area open.

At the IWC's 1964 meeting, the Norwegians reported that they had seen just eight blue whales the previous season, of which they killed four. The following year, only the Soviets found blue whales, and they reported killing 20 of them. (In fact, they killed far more, but were underreporting their illegal catches.) So this is what it had come to: blue whales in the Antarctic were not quite biologically extinct, but they were on their way, and they were certainly far too scarce to be worth the expense and effort of hunting them. The British and Dutch had already left the Antarctic entirely and begun selling off their vessels and equipment. The last shore station on South Georgia had closed forever after the 1961–62 season. In what must surely be the most belated conservation measure ever enacted, the nations of the IWC agreed in 1965 to a complete ban on killing an animal they had already ushered into oblivion. They may as well have made it illegal to hunt unicorns.

Perhaps the most surprising thing about the IWC's ineptitude was that many of its members were still in denial years after the stocks of blue whales collapsed. In 1974, the American J.L. McHugh, a former chairman of the commission, published an apologia for the IWC's performance. While admitting that the organization was imperfect, he wrote:

> Most people think that "whales are an endangered species," as I read in an article not too long ago. This view ignores the fact that, of the approximately 100 different kinds of Cetacea, less than 20 are being taken commercially, and less than 10 of these have been overharvested to the point at which they can be considered "endangered."[10]

Here was one of the IWC's most influential figures—who was not even from a country involved in commercial whaling—scoffing at criticism of the industry because it had threatened to destroy *only half* of the species it harvested, and because it had generously left alone the species that had no commercial value. The IWC had failed for four decades to protect the most magnificent animals ever to have inhabited the oceans, and its spokesman brushed this aside because the industry had not killed every last harbour porpoise. McHugh went on to argue that a total moratorium on commercial whaling "is not only irrational but unnecessary." He was wrong about that, too. By the time commercial whaling finally stopped in 1986—and this was only possible because the IWC was by then dominated by anti-whaling nations enticed to join by the United States and environmental groups—it had succeeded not only in overseeing the destruction of hundreds of thousands of whales, but also the annihilation of its own industry.

While the IWC has been justifiably criticized for its incompetence, it is now clear that even if it had set sustainable quotas, these alone may not have been enough to protect the blue whale. Illegal and unregulated whaling went on outside the IWC's influence even after the total ban on killing the species, and the full extent of it was not known until the mid-1990s.

The most notorious pirate whaler of them all was Aristotle Onassis, the Greek-born Argentinian shipping magnate who married the widowed Jacqueline Kennedy. Onassis entered the whaling business in 1950 when he converted a tanker into a factory vessel he called the *Olympic Challenger*, hired a former Nazi collaborator to manage the enterprise and announced he was going to sea with a dozen catcher boats. (To avoid even the IWC's toothless regulations, the fleet sailed under the flags of Panama and Honduras, neither of which were members of the commission.) The litany of offences committed by the *Olympic Challenger* range from killing whales out of season to submitting bogus catch data to taking undersized animals. German crew members later came forward and admitted that during the autumn of 1954 the crews killed 285 blue whales in Peruvian waters and reported them as sperm whales. Thirty-seven were under 59 feet, and therefore immature. They also confessed that during the 1954–55 season, they exaggerated their catch numbers to the Bureau of Whaling Statistics so the quota would be reached sooner. Once the other whalers left at the close of the season, the *Olympic Challenger* swooped in and killed another dozen blue whales.

As odious as the *Olympic Challenger* affair was, it paled in comparison to the illegal whaling that went on under the IWC's nose. The villain in this story is not a single pirate vessel, but rather the entire Soviet pelagic fleet, which illegally killed more than 100,000 whales—including almost 10,000 blues. While the commission could not be held responsible for the actions of an Aristotle Onassis, it must be taken to task for its failure to prevent this slaughter, which was carried out by one of its member states. Had the IWC placed international observers on all factory vessels—something its members had argued about for over a decade—the abuse almost certainly would not have occurred. Indeed, when observers were finally required in the 1972–73 season, the worst violations stopped almost immediately.

While some whaling companies had long accused others of falsifying their catches, the evidence today suggests that, by and large, most nations complied with the IWC's regulations. After the USSR dissolved in the early 1990s, however, a group of scientists that had worked on whaling vessels decided to reveal the truth about Soviet activities. They had plenty of hard evidence: although the fleets had falsified their reports to the Bureau of Whaling Statistics, they kept accurate records for their own purposes, and the numbers were appalling. Between 1947 and 1972, Soviet whalers in the southern hemisphere illegally killed a staggering 43,000 humpbacks, 21,600 sperm whales and at least 9,200 blues. The latter were almost all encountered while targeting other species in the Indian Ocean, and many were taken long after the IWC's total ban: as late as 1971–72, they killed more than 500. After 1961, the Soviets also illegally killed almost 700 blues in the North Pacific.

When these revelations first came to light in 1993, they were shocking, and they went some way toward explaining why blue whales were not recovering as quickly as expected after almost 30 years of protection. Decades of stock assessments and management plans were suddenly called into question, since they were drawn up using incomplete data. In 1998, the IWC elected to remove the USSR's catch statistics from its database, and since then a team of scientists—including the Russian whistle-blowers—has been analyzing and correcting the new data with the aim of learning the true death toll for each species. Their progress has been frustratingly slow, and even when this work is finished, the data will not be complete, because not every Soviet expedition kept careful records. The legacy of this quarter century of indiscriminate and illegal whaling, including its long-term effects on the blue whales of the southern hemisphere, will never be fully understood.

———

Judging past generations according to today's ethical standards is always dangerous. It is easy for people in the 21st century to rail against whalers for their supposed savagery, or for the recklessness that kept them firing their harpoon guns even after they must have known they were hunting the blue whale to extinction. But one should be careful to remember the context: the flensers, lemmers, engineers, mess boys, and even the gunners, were making a modest living doing extremely difficult work in an era when virtually everyone saw animals as nothing more than a natural resource for human consumption. Indeed, after the Second World War, the need for the edible fats that whale oil supplied was a genuine nutritional concern. That does not excuse the industrialists who willingly hunted a magnificent species to commercial extinction in order to enrich themselves. It does mean, however, that the wholesale condemnation of the crews who killed these whales six or seven decades ago is unfair. It would be akin to some future generation writing off our own society as barbaric because we eat meat and wear leather. Even researcher George Small, who wasn't one to pull punches, made a distinction between the whalemen and the owners of the companies that employed them. "I found that the men who killed and butchered the whales expressed, almost to a man, genuine compassion for their victims. In sharp contrast was the attitude of whaling company officials who considered whales a mere industrial raw material, like iron ore, valuable only for their contribution to production and corporate profits."[11]

What did the men aboard whaling ships really feel as they watched an indescribably large blue whale fleeing for its life, being blasted with a grenade harpoon and getting hauled up the stern slipway of a factory ship? Many did reflect on and write about these questions, giving us a window into their thinking. For the most part, the crews appear to have had a deep fascination with the animals, a respect for their extraordinary power, a desire to dispatch them quickly and humanely, but also an unquestioned belief that their job was not fundamentally different from that of a butcher preparing beef or pork.

In the 1931 edition of *Whaling in the Frozen South*, A.J. Villiers shared a typical moral viewpoint: "It is a hard life, with plenty of grueling toil; the Norwegians earn what they get. They are entitled to their reward. If they kill whales—well, others kill elephants for 'sport.'" Writing in the year of the largest blue whale hunt in history—almost 30,000 died in 1930–31—Villiers also suggested that people believed that wiping out the species was unlikely, since humans would not be so foolish as to eliminate the source of their wealth. "I am not greatly concerned about the fear of the blue whale's extinction . . . The very immensity of the industry is the whales' best guarantee against extinction. For it naturally follows that, as soon the whales begin to thin out at all they will have to be left alone." Villiers was far from alone in this naive opinion. "One would not like to see the poor old whale extincted," he concluded. "He is a harmless beast."[12]

More than three decades later, when the poor old blue whale had indeed nearly been extincted, the British chemist Christopher Ash chimed in with his own thoughts about the ethics of the industry. Ash, who spent almost 20 years aboard whaling ships, was not insensitive to the majesty of the blue whale. "It is a beautiful creature when disporting itself at its leisure," he mused in *Whaler's Eye*, "and wonderful to see when running for its life."[13] But neither was he under any illusion that its real value was anything other than economic: "It is cheerful to walk along one of the main alleyways beside the bone boilers when the factory is rumbling away at 'full cook,' and every sweating face has a smile. 'I smell money cooking,' is how the Norwegians aptly described it."[14] At the end of his book, Ash constructed a Platonic dialogue to refute the objections of a hypothetical critic of the whaling industry:

Q But tell me, Ash, how do you feel when you are in a catcher, and see a whale chased and killed.

A The chase is very exciting, and I desperately want the gunner to catch the whale, even while admiring the strength and speed of the whale, with its cunning twists and turns . . . Seeing the death on the surface when the whale wearily rolls over as it gives up its life is very moving. The sight of so much strength and beauty disappearing cannot fail to arouse pity.

Q Yet you keep whaling?

A Yes, the pity which I feel in the antarctic is better than the distaste experienced after seeing the sordid details of a slaughterhouse. I have accepted both as being necessary.[15]

Gibbie Fraser, recalling his years aboard catcher boats in the Antarctic, stresses that the gunners always tried to bring about a swift death. "Nobody liked to see the whale suffer. It wasn't nice to see the killer harpoon go in and watch the body shudder when the grenade exploded inside. When you got one that was shot badly, then everybody did their utmost to bring about its demise as quickly as possible. But there was comfort on the other side—this was your bread and butter. This is what you had come for, to make your money on, so you had to get them." On the question of how long it took a large whale to die, Ash was either dishonest or he hid belowdecks whenever the gunner took aim. He wrote that "the death struggle cannot be more than a few minutes,"[16] but this is certainly untrue. Fraser himself saw battles that lasted much longer. "It usually took just one killer harpoon for a big baleen whale—they were much easier to kill than sperm whales. But the agony might go on for half an hour. It seemed that sometimes you just could not find the vital organs to kill them. I remember one or two occasions where a whale was badly shot and it took out a whole other line. The winch was very powerful and very slow, and you had to heave all the line back again and coil it neatly in the hold of the ship. This took a long time, and all the while this whale was thrashing at the far end."

Whaling captain W.R.D. McLaughlin dispelled any notion that the animal did not suffer terribly on the end of that line. "How few people realize the despairing fight it puts up before dying of convulsions with an explosive shrapnel grenade tearing at its vitals. It has to be seen to be believed."[17] The fact was, however, that when McLaughlin wrote those words in 1962, almost no one outside of the industry *had* seen the spectacle of a blue whale in its death throes. Antarctic whaling expeditions rarely carried journalists, and even if reporters had ventured into the ice and emerged with stories and photos, readers in Europe and North America would not necessarily have been spurred to action. The environmental movement in the west—traditionally traced to Rachel Carson's *Silent Spring*—was still embryonic. Greenpeace and its harpoon-dodging hippies came a decade too late for blue whales. When the hunting of the species stopped, it was because the animals had disappeared, not because of any public outcry or change in sensibility.

Then something remarkable happened—a dramatic about-face. Around 1970, the blue whale was transformed in short order from the ultimate quarry of hunters to a species with exalted status. After hundreds of thousands of animals had been turned into soap and margarine without raising more than a whisper of concern, the blue whale suddenly became a conservation icon.

What makes the blue whale's ascendancy so unlikely is that it happened even though almost no one, including many whalers and most marine biologists, had ever seen a blue whale. In the era before whale-watching tours and Animal Planet, large baleen whales could not have been farther removed from people's daily experiences. Photographs of blues in the wild were rare, and most were taken aboard whaling ships moments before the animal took a harpoon in the back. These whales were not adorably cute like the seal pups that would later preoccupy animal rights activists; they were nothing more than vague humps protruding from the sea. And yet, in the words of whale biologist Steven Katona, "perhaps more than any single animal, the blue whale stimulated the resurgence of public interest in ecology."[18] How did the enigmatic and almost invisible blue whale capture so many human hearts?

Katona is in a position to answer that question because he watched the transformation happen. He was doing graduate work in biology in the late 1960s, and completed his PhD at Harvard in 1971. The following year he was a founding faculty member at the College of the Atlantic in Bar Harbor, Maine, and spent much of the next three decades studying whales in New England and beyond. Katona remembers the sharp increase in awareness that occurred after the American biologists Roger Payne and Scott McVay recorded the ethereal sounds made by humpback whales off Bermuda in 1967. Four years later, Payne released an LP titled *Songs of the Humpback Whale*, which became an unexpected bestseller and generated enormous interest in the mysterious and vulnerable creatures that created such music. "They are eerie, haunting sounds, and it was altogether too easy to hear them as a cry for help," wrote Richard Ellis. "Not only were the whales on their way to extinction, they were singing their own dirge."[19] We know now that humpbacks and blue whales make entirely different vocalizations, but only specialists were aware of this at the time, and so all whale species were suddenly in the ears and on the minds of the public like never before.

As the scientific understanding of whale sounds deepened, Payne co-authored a landmark paper in 1971 that again shoved the blue whale to the forefront of the environmental movement. The U.S. Navy had for decades been listening in the oceans using highly classified arrays of hydrophones. These instruments were designed to detect enemy submarines, but they also picked up low-frequency pulses that were believed to be made by baleen whales. Along with a colleague, Douglas Webb, Payne argued convincingly that the animals making these mostly infrasonic calls—primarily fin whales, but blues as well—were using them to communicate with one another. The calls were extraordinarily powerful, and their low frequencies would allow them to travel vast distances, perhaps even across ocean basins. Hardly any non-scientists read the technical paper, of course, but Payne discussed his idea publicly and it stirred strong emotions. "Roger is a magnificent speaker," Katona says, "and he spoke very eloquently about how these sounds could penetrate throughout the ocean." People were moved by the idea of a blue whale calling to others of its species, perhaps in vain, since the one thing most people did know about blue whales was that they had been hunted to near extinction. "It was an image of loneliness in the ocean. A blue whale's call spreading out for thousands of miles, but with no other whale to answer, was a very powerful image of the last moments of the largest species ever to have lived on earth."

Although the new discoveries involved humpback and fin whales more specifically than blues, these distinctions were mostly lost on nature lovers in the early 1970s. "You have to remember that almost no one had seen any whales," Katona says. "People didn't know the difference between a humpback and a blue whale in any real sense. There wasn't very much popular imagery, and what imagery there was wasn't very accurate." Even today, fin whales, despite being the second-largest creature in the ocean, swim largely under the public's radar. Humpbacks have become much more recognizable because they're commonly encountered off New England, Hawaii and other whale-watching hot spots, and they're the ones seen breaching and fin-slapping in photos and videos. During this period of growing concern about whales, however, no one had any reason to identify more strongly with humpbacks. Instead, they gravitated toward the species that commanded all the superlatives. "The blue whale became the one that was exceptionally special," says Katona, "because it was the most endangered, and because it was the biggest animal on the planet."

In some ways, the fact that most people would never see a blue whale made it the perfect symbol for the budding environmental movement. It was, paradoxically, both universally recognized and completely mysterious: everyone knew that blue whales were critically endangered, though virtually nothing was known about their behaviour, intelligence or any other characteristic that people ascribe to animals they admire. This allowed a new generation to create a mythology around the species. Won over by the likes of Flipper the dolphin and Shamu the orca, people thought of whales as compassionate and deeply intelligent. Some of the earliest whale-watchers, approached by curious grey whales in Mexican lagoons, described their encounters "as an attempt by the whales to communicate with us, to express their forgiveness for the havoc we wreaked on their brethren."[20] What could be more compelling than a beautiful animal with an emotional repertoire that rivalled our own? Taking poetic licence to the extreme, some authors wrote mawkish fiction that anthropomorphized the animals. One begins: "Musco whales a boom of delight, as happy Blue whales do."[21]

One book from this era was particularly influential. Rather than ascribing human emotions to blue whales, it described everything science knew about the species—admittedly, that wasn't much—and related in bloody detail how whalers had reduced it to perhaps a couple of hundred individuals. The book was George Small's *The Blue Whale*. Published in 1971, it won a prestigious National Book

Award the following year, and for the first time it made the blue whale's plight widely known. Small was given to bouts of melodrama, and much of the science he reported has turned out to be inaccurate, but he was a diligent researcher. Like a prosecuting attorney, he documented the misdeeds of whalers and the IWC and wrote moving passages that appealed to readers who were learning about the animal for the first time. In his final chapter, Small refers to "the tragedy of the passing of the blue whale at the hand of man,"[22] using the past tense, as though the species had already vanished. He writes off the animal as doomed in the Antarctic and the North Pacific, though he holds out a tiny flicker of hope: "If by some miracle the blue whale does survive anywhere, it will be in the North Atlantic where no pelagic whaling has been carried on for several decades. Even there, however, the chances for survival of the blue whale are virtually nil."[23]

Fortunately, the death sentence was premature. Just five years after Small's book was published, a 24-year-old American biologist named Richard Sears spotted his first blue whale in Quebec's Gulf of St. Lawrence, an arm of the North Atlantic. It was a meeting that would alter the direction of Sears's life—and the study of blue whales—forever.

FIVE

ST. LAWRENCE BLUES

I T IS 1:15 p.m., THREE HOURS AFTER LEAVING the marina at Portneuf-sur-Mer, Quebec, in the St. Lawrence estuary, when Richard Sears spots his first blue whale of the day. As he grabs his binoculars, the whale surfaces and sends a plume of vapour 25 or 30 feet into the air. At the wheel of their small boat, his colleague Andrea Bendlin picks up her slate and glances at the boat's instruments, noting the GPS coordinates and the depth of the water, which is almost 200 feet. "Big dorsal fin," says Sears, peering through the binoculars. "Balanced pigmentation. It might be Étrier." The 26-year-old Bendlin, in her second season with the research team, smiles as Sears moves to the bow to get his camera, still in awe at how the biologist is able to recognize individual whales from a hundred yards off in a moving boat. "Yup, it's Étrier," says Sears, focusing his telephoto lens and snapping a couple of digital photos. "Two pictures, both left side," he calls to Bendlin, who writes down the information. Étrier is more formally known as B035 because he was the thirty-fifth blue whale to be identified by Sears in the St. Lawrence. His nickname, French for *stirrup*, comes from a distinctive shape in the mottling on his back. Bendlin will hold off naming Étrier in her notes until the ID is confirmed by comparing these photos with the database later that night, but Sears isn't wrong very often. "I probably recognize these whales more easily than I recognize people I know," he says.

Just before ten that morning, Sears and Bendlin had started the twin 90-horsepower outboards and set off in their 23-foot rigid-hulled inflatable, nicknamed *Baleine Bleu*. This type of boat has become the standard vehicle of whale biologists, who like its manoeuvrability, safety and fuel economy. The beautifully clear September sky is streaked with contrails, and while it is warm onshore, the temperature plummets dramatically on the water, so Bendlin wears gloves and a wool hat, while Sears sports his trademark Boston Red Sox cap. The night before, the biologist had been chatting with some policemen at a truck stop in the small town of Forestville. The cops told him they had flown a helicopter over nearby Pointe à Michel that day and had seen a number of whales, so Sears and Bendlin wanted to check it out. Local people, including whale-watch operators, ship captains and fishermen, often keep Sears abreast of what's happening, though a few are suspicious of the guy in the red survival suit. One whale-watch captain started an absurd rumour that Sears was putting satellite tags on blue whales so Japanese whalers could track and kill them. Others reacted angrily when a local magazine misquoted him as saying there were few large whales in the area; they believed he was trying to discourage tourists. Most encounters are more friendly, though, and shortly after leaving the marina that morning, Sears tries again. He pulls alongside a fishing boat and calls to one of the men on deck. "*Bonjour! Avez-vous vu des baleines ici?*" Have you seen any whales around here? Sears speaks perfect French, though his accent is distinctly lacking the Québécois twang. The fisherman says he just saw four or five *petits rorquals*—minke whales—a few miles off Pointe à Michel, corroborating the policemen's story. But an hour of cruising turns up nothing beyond a harbour porpoise and a couple of grey seals. Sears has to rely on his own knowledge and experience to find Étrier a couple of hours later.

Once the big male has left, Bendlin pops open the cooler and tosses Sears a sandwich and an apple. It's a quick lunch, as Sears does not like to waste time on the water. His crewmates quickly learn to take advantage of these brief breaks to fill their stomachs and empty their bladders, because if whales show up, they won't get another opportunity. Within a few minutes, the scientists are heading farther upriver to Les Escoumins, usually a reliable place to find blue whales this time of year. Sure enough, less than half an hour after arriving, they recognize Jawbreaker, or B246, who is one of the easier whales to identify: she has a distinctly whitish dorsal fin, and on her terminal dives she raises her enormous tail flukes above the water, a behaviour common in right whales, humpbacks and sperm whales, but unusual among blues. Jawbreaker has a strong personality—males regularly approach her, but she usually tolerates them for only a short time before heading off on her own. Today there are two other whales in the vicinity—Sears later identifies them as Zaffre (B101), a female, and Pewter (B185), a male—but they do not seem to be associating with one another, concentrating instead on their meals. When Pewter disappears, a flock of kittiwakes and Bonaparte's gulls, both krill-eating birds, heads immediately for his footprint, the smooth, round patch formed on the water by the upstroke of the diving whale's tail.

By about five in the afternoon, Jawbreaker has moved off and Sears and Bendlin spot two other blues swimming side by side. Over the years, Sears has learned that when you see a pair of blue whales together for a prolonged period, it's almost always a female being followed by a male. Understanding the behaviour of blue whale pairs has been one of Sears's main research goals, so he's eager to identify both individuals. As Bendlin moves the boat closer, he photographs the right side of the whale in front. "Three pictures of the lead animal," he calls to her. Then another trio of clicks. "Three of the trailing one." The pair has been surfacing together at seven- and eight-minute intervals, and Sears has now recognized them as Pulsar (B036) and Éperon (B227). Surprisingly, both are males. Suddenly the two whales come up just a few yards from the boat, their exhalations like small explosions, each followed by a shorter sucking sound as they take in a gulp of air that could probably fill a small blimp. Without warning, Éperon takes a sharp turn to the right and heads straight for the boat, which suddenly feels pathetically small, and his blue-green form glides directly beneath. The whale is unimaginably big. Its head is perhaps 15 or 20 feet across—almost as wide as the boat is long—with a central ridge that forms a splash guard in front of a pair of massive blowholes, now closed up tight. Before all 75 feet of his body have completely cleared the boat he surfaces again, the blowholes opening up and blasting the scientists with a salty spray. Sears protects the lens of his camera and then lets out a whoop of delight. Even after 30 years in the presence of these magnificent animals, he finds an encounter like this unforgettable. As one veteran scientist put it, "There is no whaler and no whale biologist, no matter how experienced, who is so jaded that his heart does not race at the sight of a blue whale."[1]

It is probably fair to say that Richard Sears has spent more time in the presence of blue whales than any other human being. For three decades, he has followed the ocean's largest animals, photographing them, identifying individuals, observing their movements and behaviour, and trying to unlock the secrets of their lives. Most winters Sears travels to the Sea of Cortez, which separates the Baja Peninsula from mainland Mexico, to observe the blue whales that arrive there in January and February, often with calves in tow. He makes regular excursions to the waters off Iceland and the Azores to study the blues of the eastern North Atlantic. In 1997, he journeyed to South Georgia as part of the first marine mammal survey in that area since the end of whaling. But it is in the Gulf of St. Lawrence and its estuary that Sears spends most of the year, studying its population of 400 or so blue whales, many of whom he knows by name. "One of the things I always dream about is being able to live to be 80-odd years old and seeing some of the animals that I had known in my twenties. That would be pretty amazing."

Strong and fit, with the rugged features of a man who has spent much of his life on the water, Sears was born in Paris in 1952 to an American father and a French mother and grew up speaking both

Richard Sears spotted his first blue whale in the St. Lawrence in 1976. Three years later, he started the first organization devoted to long-term research on blues.

English and French. He has mariners on both sides of the family—his father was in the U.S. Navy and his maternal great-grandfather captained a vessel—and he suspects he may even have New England whalers in his ancestral tree. He was the type of kid who would come home dirty because he had been collecting caterpillars or exploring the woods. In his catechism lessons he challenged the nuns who told him to believe things that clashed with his own observations, much like the way he questions the received wisdom in whale biology today. He combined his interest in wildlife and his love of the sea by earning a degree in biology and naval history at Nasson College in Springvale, Maine, graduating in 1975. The following year he was hired by Woods Hole Oceanographic Institution in Massachusetts, who sent him to Matamek Research Station in eastern Quebec, on the north shore of the St. Lawrence. Working under a researcher named John Gibson, the 20-something Sears set about studying the behaviour and distribution of salmon. It wasn't long, however, before he found himself distracted by the large whales that visited the station. "I just observed from shore, morning, noon and night at mealtimes, and if I saw any whales out in the bay, I noted them. Eventually I convinced John to let me go out in the inflatable. They were mostly fin whales, but that's also where I saw my first blue whales, and I'll never forget how big the blow-holes looked. At one point I had the whole station going out, including John. After four days of chasing whales he looked at me and said, 'Richard, do you think we might get back to salmon research?'"

Sears continued studying salmon, but he kept up his informal whale-watching, and that October he made a discovery that would change the direction of his career. "I saw some right whales off Matamek, and that was the first time right whales had been seen alive in the St. Lawrence for over a hundred years. John pushed me to write a little note for a journal, which I did, and that made me notorious with certain elements." It turned out that the veteran whale researchers in the area resented being scooped. "They didn't want to believe that this little schmuck working on salmon had seen some right whales, and I was grilled as though I were being talked to by the CIA." That was Sears's first contribution to the understanding of whales in the St. Lawrence. There would be many more.

Sears completed his salmon work in February 1977, then returned to New England and spent parts of the next three summers at Mount Desert Rock, Maine, helping to conduct a census of marine mammals. While there he worked with Steven Katona, who was one of the pioneers of a relatively new technique at the time: using close-up photographs to identify individual humpback whales, and then tracking these whales over time. He didn't realize it yet, but getting to know Katona and learning to ID whales from photographs would later help Sears become a pioneer in his own right.

Although New England was his home, Sears has always been driven to do original research, seeking to answer questions no one else has tackled, and his curiosity drew him back to the whales he had first seen at Matamek. So he returned to Quebec in 1979 and established the Mingan Island Cetacean Study (MICS), the first organization devoted to studying whales in the St. Lawrence, and the first in the world to undertake long-term research on blue whales.

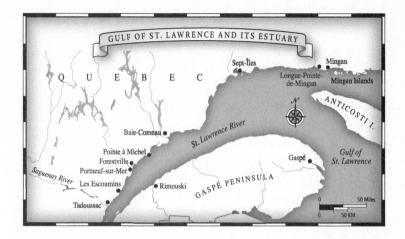

The MICS research station, which includes a small museum, is in the village of Longue-Pointe-de-Mingan, a short boat ride from Quebec's Mingan archipelago, a group of about 40 large limestone islands and hundreds of islets, spread out between the north shore of the St. Lawrence and Anticosti Island. These islands and surrounding waters support hundreds of animals, including Atlantic puffins, eider ducks and countless other birds, as well as three seal species. At one time the sea here was brimming with cod, though these have all but disappeared after centuries of overfishing. Today one of Mingan's main attractions is its whales: minkes, fin whales and humpbacks are regular visitors. Blue whales were once common here, too, though in the early 1990s they mysteriously disappeared—more about that later.

Sears's team arrives at the station each year around the end of May and as the season progresses they cross the St. Lawrence to the waters off Gaspé, or head east along the north shore of the Gulf, travelling as far as the Strait of Belle Isle between Newfoundland and Labrador. In late August and September, they move west into the estuary, where the fresh water from the St. Lawrence River meets the salt water from the Gulf. The estuary's average depth is about 650 feet, but features on the seafloor cause the colder, nutrient-rich salt water to be pushed toward the surface in some places. These so-called upwelling areas support some of the densest concentrations of krill in the northwest Atlantic, which is what attracts the baleen whales. During the late summer and early fall, the whales move as far upstream as Tadoussac, a popular tourist destination. The estuary is one of the only places in the world where blue whales, more often seen well offshore, can regularly be spotted from land; some of the homes along the north shore have decks where people can barbecue and sip beer while watching the spouts of the planet's largest inhabitants. By late fall, most of the blue whales have left the estuary, although some will stay as late as January if there is little ice, and by winter they're rarely seen in the Gulf. Where do they go when they leave? That is one of the mysteries Sears would like to solve.

MICS has an annual budget of between Cdn$150,000 and $220,000, and unlike many research groups that rely heavily on government contracts, it gets a good deal of its funding from ecotourism. Each year 30 or 35 people from around the world pay about $2,000 a week to join Sears and his colleagues in either Quebec or Mexico, riding along in small boats and getting a first-hand look at the daily work of whale researchers. It's not a model that appeals to every scientist—some cringe at the idea of ignorant whale gawkers getting in the way of their work—but Sears enjoys "tenderizing the tourists" who are accustomed to more comfortable vessels. He remembers one group going home to write an article advising others on how to prepare for the session. Their suggestions included bouncing on a couch with rocks in their pockets while someone blasts them with a firehose.

MICS keeps its costs low by relying on a rotating crop of a dozen or more unpaid interns each summer. There are also six to eight team members who get room and board and a small paycheque. Sears says one of his prime motivations, in addition to his own research, is giving these students an opportunity to demonstrate their abilities in a nonacademic environment. Never having pursued a graduate degree himself, and far more comfortable in an inflatable than at a marine mammal conference, Sears looks for interns who share his curiosity and attraction to research no one else is doing. He interviews the applicants rigorously and looks not only at their experience in boats or working with whales, but also at their enthusiasm and willingness to take risks. "What always surprises me is that we never have to advertise for people to come and work at the station with us. You always find young kids with the same passion, the same fire we had when we started. These kids join us for anywhere from one to ten years, sacrificing themselves for a while before waking up and realizing that the guy who runs the place is a nutcase and that they have to go and make a living."

For Sears, studying whales has never been just about making a living. "When I started doing this kind of work, all of us were having a romantic adventure, which is what we were interested in as well as the science. That's an advantage to being in at the beginning of something, before it gets more civilized." The whale biologists of the 1970s, who began the first long-term studies of live cetaceans in their natural habitat, concerned themselves with simply getting out in boats and observing and recording what they saw. They were natural historians who didn't think in terms of population models, statistical analysis and other sophisticated tools that scientists use today. Now when Sears attends conferences that once drew a few dozen people who all knew one another, he sees hundreds of young scientists with different motivations. "They're very serious, and very exact in what they do. A lot less wild than we were. That's good, and that's always the way things seem to happen. You have the adventurous and entrepreneurial types that start things, and then others come along and build on it, and probably do a better job." Sometimes this difference in mindset does create tension between the mentor and his proteges, though. Sears likes to needle some members of his team for writing boring field notes that include only lists of observed animals and coded behaviours. They respond by telling him that it's not much use to science to describe a beautiful sunset in the research log. Sears shakes his head at the criticism. "Logbooks are meant to be *written*. I don't want to know about sunsets, but I do want people to write their impressions of the animals they see. Make sure the data is there, but also add to it. It helps you remember things about the day."

The scientific literature on blue whales includes surprisingly few articles that name Sears as the lead author. However, many papers by other scientists rely on his work indirectly; it's not unusual for an author to describe something about blue whale behaviour and cite "Sears, personal communication." This shorthand means that the author, unable to rely on published data, simply phoned or e-mailed Sears to ask for his insights, and that these were authoritative enough to include in a peer-reviewed journal. Why hasn't he written more papers himself after so many years of research? One reason is that he has evaded the publish-or-perish environment of academia—no university department head ever nags him to generate new articles. Another factor is simply his professional restlessness: by the time members of the MICS team have analyzed the data and arrived at an interpretation, Sears has often moved on to another question and he doesn't always make the time to write up something that seems like old news to him. Finally, so much of what Sears has learned by observing blue whales is subtle and hard to quantify in a scientific paper. It's not so much data as it is wisdom.

When Sears began his work in the St. Lawrence, knowledge about blue whales in the western North Atlantic was minimal. Scientists had noted blues off the southern coast of Newfoundland, and the few researchers in the estuary were working mainly on other species. In 1979, the year Sears returned to Quebec, some of these scientists conducted a three-month survey of baleen whales in the estuary, but they did it all from shore with binoculars and spotting scopes. This resulted in little useful data, Sears says. "They knew blue whales were in the St. Lawrence, but that was about it. They didn't know much about their movements or how many there were." What was needed, he knew, was a means of identifying individual whales.

The ability to recognize individuals is an essential tool for studying mammals; it is the basis for estimating population sizes and understanding a species' distribution and movement patterns. In the late 1970s, however, the approach was just catching on among whale researchers. Roger Payne first tried it early in the decade with right whales off Argentina. He took photos of the animals whenever they surfaced and was able to identify each whale by its unique pattern of callosities, the rough patches of whitish skin on their otherwise black bodies. Later researchers noted individual grey whales off British Columbia by their distinctive markings and the knuckle-like ridges along their backs. Others identified orcas by their dorsal fins and the shape of their greyish "saddles." Sears's colleague, Steven Katona, assembled a catalogue of humpbacks in New England, distinguishing each whale by the markings on the ventral side of its tail flukes. But no one had made a concerted effort to apply these techniques to blue whales, and some researchers thought Sears was crazy to even try.

Blue whales do present some challenges when it comes to photo-ID. Unlike right whales and grey whales, they have smooth skin that is largely devoid of callosities, barnacles and whale lice. Blues also have extremely small dorsal fins relative to their enormous bodies, and while the size, shape and colour of these vary, the differences are not as dramatic as they are among killer whales. Humpbacks generally lift their tails out of the water when they dive, making their flukes easy to photograph, but only about one in seven blue whales is a "fluker." What, then, could Sears use to identify individuals? His solution was to look at the patches of light-coloured pigmentation along their backs. "It made perfect sense to me—why wouldn't these marks be useful the way a humpback's ventral fluke pattern is? It was not really that much of a leap." Sears realized that every individual has a uniquely mottled pattern that remains consistent as the whale ages; blue whales, like leopards, cannot change their spots. The patterns also show a surprising amount of bilateral symmetry, meaning it is unusual for a whale to be heavily mottled on one side but lightly patterned on the other. Sears quickly learned to keep the sun at his back to achieve the right lighting, and to use the dorsal fin as a reference point, snapping his photos as the animal arched its back before diving. Working from inflatables, he and his team photographed both the right and left sides of every whale they approached, as well as the tail flukes whenever that was possible, and noted any distinctive scars. They labelled and dated the photos, made notes on ledger sheets and organized them in file folders. Whenever they obtained a new photo, they compared it with the others to see whether it was an unknown animal or one they had previously seen. Sears was on his way to compiling a catalogue of blue whales in the St. Lawrence, the first of its kind in the world.

For several years, some old-school researchers remained skeptical about identifying blue whales from photographs, and it wasn't until around 1987 that the technique gained widespread acceptance. That year Sears wrote a journal article about how he had identified blue whales in the Sea of Cortez, and he published his first St. Lawrence catalogue so others could see how one was assembled. The following year, the International Whaling Commission held a workshop in La Jolla, California, on nonlethal research techniques. (After the 1986 moratorium on commercial whaling, scientists had far fewer carcasses to study.) At this meeting, attended by scientists from around the world, Sears demonstrated how he had identified 203 individual blue whales in the St. Lawrence. "That changed everything," Sears remembers. "Before that, people in New England had accepted photo-ID, and those of us doing the work were very confident about it, but a lot of the old buggers at the IWC still thought it was stupid."

Since then Sears has become so familiar with his catalogue of blue whales, which has grown to more than 400 individuals in eastern Canada, New England and western Greenland, that he can usually match a photo, or establish that it is a new animal, in about 15 minutes. "Everyone else takes about three hours," jokes Andrea Bendlin, exaggerating only a little. (In 2004, Sears photographed a whale that he couldn't identify on sight, "but something twigged in my memory." It turned out to be a whale he had last seen during an aerial survey 22 *years before*.) All the old black-and-white prints have now been scanned and the new images are digital, but the matching is still done the old-fashioned way; no one has yet created software that can compare photos of whales taken from different distances and angles. Sears's team has come up with a system of classifying pigmentation patterns so they don't have to sift through the entire catalogue every time, however. If the blue whale's mottling is more or less evenly distributed between the blowholes and the tail, they classify it as *balanced*. If the pattern gets more or less dense along its body length it's *merging*, and if the variation looks more like a series of horizontal bands it's *tiered*.

In addition to photos, the MICS database contains everything known about each individual, including its sex, the dates and locations of every sighting, and whether it has ever been spotted in a pair or larger group. About a third of the whales in the eastern Canadian catalogue have been seen only once. However, it has given Sears a good understanding of how and where blue whales move from the time they show up in the Gulf in spring until their departure in late fall or winter. "We know where we can expect to see them at different times of year," he says. "We know that if you go to the tip of Gaspé in May you're likely to see blue whales, and some of those animals are likely to stay there until July, then they'll start moving, most likely upriver." But blue whales are complex animals that continue to surprise Sears with their seasonal movements. "This thing about wholesale migrations drives me nuts: 'OK, it's October, so everything has migrated automatically.' That's not how it happens. Even with humpbacks, certainly many more of them migrate, but there are some that don't: females that have given birth the year before, some that are past reproduction, some that are just tired and fed up. With blue whales there might be some migration, but there are all kinds of scenarios."

One of the more remarkable scenarios has been the disappearance of blue whales from Mingan, the very place where Sears first encountered the species in the 1970s. They were regulars there until the early 1990s, when they vanished "as if aliens came and beamed them up." Some of the animals that had been identified in that area have never been seen since. What could have happened? Sears believes their disappearance might be related to the collapse of the cod fishery, which occurred right around that time. (After overfishing reduced stocks to historically low levels, the Canadian government declared a moratorium on cod fishing in 1992.) Mature cod are voracious predators of other fish, and with fewer of them around, other species such as capelin and sand lance are now thriving. "Maybe the capelin and sand lance have been hammering the krill, so now there's no reason for blue whales to come to Mingan because there's too much competition," Sears speculates. Even after decades of asking questions like these, the answers are coming together only slowly. "The longer I do this, the less I know. The whales have to give you this stuff. You can't just take it."

Spend any time with Richard Sears and you'll notice that underneath the confident, sometimes brusque exterior—not to mention his notorious intensity on the water—is a profound humility around the animals he has spent his life studying. He has never tried to pretend that he fully understands blue whales. "Everyone always asks me how intelligent these animals are. I don't know. I think humans have a difficult enough time evaluating our own intelligence. Only after the Human Genome Project did we find out we barely have twice as many genes as a fruit fly. I think that's great for humans—it's a great comeuppance. Yet we still don't seem to get it."

The International Whaling Commission, still using boundaries it drew up in 1991, treats the blue whales of the North Atlantic as a single stock. For more than 50 years, however, some whalers and researchers have believed that these whales make up at least two discrete groups. The most recent evidence supports this latter view, suggesting a western population that includes eastern Canada, New England and western Greenland, and an eastern population that ranges from the waters off Iceland, south past the Azores (the island group some 930 miles off Portugal) and ranging at least as far as

northwestern Africa. Whether there are subgroups within these two populations, or whether there is any mixing between blue whales from opposite sides of the Atlantic, is so far unknown, and it is part of what long-term photo-ID studies can determine.

The blue whales of the North Atlantic and its adjacent Arctic waters were the first in the world to be targeted by whalers. Indeed, these giants were the ones that made Svend Føyn give up the seal business. Blues were vigorously hunted off Finnmark, in northern Norway, until the early 1900s, and they have rarely been seen there since. Farther west, in the waters off Iceland, however, blue whales have been known since the earliest days of whaling. In the first half of the 20th century, fleets also took blue whales in the Davis Strait, between western Greenland and Baffin Island, as did whalers from shore stations in Newfoundland and Labrador. Blue whales were also known to travel through the Strait of Belle Isle into the

Gulf of St. Lawrence, and the whalers followed them. In 1911, the Norwegian-Canadian Whaling Company opened its doors in Sept-Îles, less than two hours' drive from where Richard Sears's research station now sits, and operated for five seasons. The records are not clear on just how many blue whales died there; the station took about 400 whales in total, and perhaps a third of those were blues. If blue whales can indeed live for 80 years or more, as many scientists believe, it is possible that the offspring of some of these slaughtered whales are today in Sears's database.

A whaling station in Hawke's Harbour, Labrador, took a few blue whales as late as 1951, but by then blues were extremely rare in the North Atlantic and the species was commercially irrelevant. Following the usual pattern, the whaling nations could now agree to protect the species, which the IWC did in 1955, though Iceland objected to the ban until 1960 lest any blue whales be missed. An estimated 1,500 blues were killed in eastern Canadian waters from the advent of modern whaling until 1915, and from the 1920s until the 1950s at least another 1,300 were killed throughout the North Atlantic. That's less than 1 percent of the 330,000 that died in the Antarctic, but the North Atlantic population never approached that of the Southern Ocean. There are not enough data to estimate the pre-whaling population. Even today there is no scientifically valid census, but the western group may not be much larger than the 400 or so whales identified by MICS, and there are probably upwards of 1,000 off Iceland. Too little is known about the other areas in the North Atlantic to make a useful guess.

In the mid-1990s, Richard Sears realized that if he could extend his photo-ID work to areas outside the estuary, he might be able to solve one of the enduring mysteries of the St. Lawrence blues: where do they go in winter? The seasonal movements of blue whales, in fact, are not clear in any of the world's oceans. In the classic model of baleen whale migration, animals in both hemispheres move to higher latitudes to feed during their respective summers. In the fall, they make for tropical and subtropical waters to give birth and to nurse their young. Grey whales are the tidiest example;

they leave their feeding grounds in the Bering Sea between Alaska and Siberia in October and travel some 6,000 miles down the coast of North America to Baja. Between December and March, females give birth and nurse their calves in the warm Mexican lagoons, while single animals look for mates. By April, most of the whales travel northward again to the food-rich Arctic waters, feeding only occasionally during their travels, and living off the fat stores in their blubber. Most humpbacks also undertake migrations from summer feeding grounds to birthing and mating areas at lower latitudes. About 4,500 of them travel from Alaska to Hawaii beginning around October, and then return about May.

One of the nagging questions about baleen whale migrations is *why* they travel to low-latitude regions, which typically have fewer resources than colder waters. Birds and terrestrial animals, quite sensibly, tend to migrate to areas where they expect food to be more abundant, not less. Biologists have suggested that whales leave the polar regions when temperatures drop because they would expend more energy keeping warm than they would swimming to more temperate waters. More recent research, however, has shown that blue whales can do just fine in frigid seas. Others believe that mothers and calves need calmer waters to thrive, or that the animals are driven to protect their young from orcas, which are more common at high latitudes. All of these ideas are controversial.

Whatever the reasons for these seasonal movements, scientists have long understood that blues do not follow regular patterns like greys and humpbacks. Clearly there are krill hot spots—such as the St. Lawrence estuary—where blue whales congregate every summer, but Sears's photo-ID work has showed that they aren't necessarily the same whales. One might visit the area for a year or two, then go AWOL for a decade before Sears sees it again. Many are seen and photographed only once. And while most blue whales begin to leave the estuary in autumn, no one knows their winter destination, or, more likely, destinations.

It didn't take long for Sears to fit some pieces into the puzzle, at least when it comes to the whales' summer range. Early on he matched a few blue whales from the St. Lawrence with sightings off Nova Scotia and New England. One whale was spotted in 1980 on the Scotian Shelf, and then resighted in the St. Lawrence estuary in 1983 and 1985. Another was photographed by whale-watchers in the Gulf of Maine in 1987, and when Sears looked at the shot he realized it was B125, a whale he had seen off Mingan in the two previous years. Shortly afterwards, Sears made his first match between the St. Lawrence and the Davis Strait. A colleague working off western Greenland photographed a blue whale in 1988 and then saw the same one again almost exactly a year later. When he sent the photos to Sears, they matched B156, a male that had been seen in Mingan in 1984 and 1985.

Every single one of these sightings, from the Gulf of Maine to Greenland, came in August, however, so they had nothing to do with seasonal migrations. What they demonstrated, however, was that the blue whales of the western North Atlantic are far more wide-ranging in summer than anyone knew. They seem to choose an area where they can expect to find food, and if they're not satisfied, they move to the next site on their list, which may be hundreds of miles away. The whales don't all use the same strategy, which is why the visitor list in the St. Lawrence changes every year. What might influence their choices, however, is anyone's guess.

As for where the St. Lawrence blues go to mate and calve, that remains unknown, too; none of the whales identified in the Gulf or the estuary in summer has ever been seen outside the St. Lawrence in winter. Sears suspects some head to the continental shelf, a couple of hundred miles offshore of the Maritime provinces. Some may go farther east to the Grand Banks and the Flemish Cap off Newfoundland—navy hydrophones have detected blue whales in these waters during the winter—while others may swim south to Bermuda or beyond. Over the years, western North Atlantic blues have turned up in some unexpected places. One was stranded on a New Jersey beach in 1916, while another swam into the Panama Canal from the Atlantic side in 1922, where it was considered a danger to shipping and shot to death. Two were stranded in the Gulf of Mexico. These were all highly unusual events, though, and it is doubtful that blue whales spend any significant time near the Atlantic coast of the United States or Central America.

If it seems surprising that the oceans' largest creatures can simply disappear from human view for months, one has to remember that most blue whale research in this region is done from small boats with a very limited range, while blue whales spend much of their time in the open sea and can easily cover 60 miles in 24 hours. "What we humans see in the ocean is usually very close to shore," Sears says, "and when you look at the vastness of what's left, it's evident that we're not going to get it all. I've had colleagues come out with us and act surprised that we're 25 to 30 miles offshore—they look at me and go, 'You guys do this all the time?' And I say, 'Sure, that's where the whales are.' But we're still pathetically close to shore, and we haven't even come close to where many of these animals may be—out in the Grand Banks and the Flemish Cap and all that. We need to find out if there are blue whales out there that only rarely come into the St. Lawrence. That may be why about a third of the whales in our catalogue have only been seen once." One of Sears's greatest frustrations is that he has been unable to mount an expedition into offshore waters east of Newfoundland. Such a research cruise, which would be enormously expensive, would likely tell him whether the western North Atlantic blue whale population is significantly bigger than the 400 he's seen in the St. Lawrence.

In search of the bigger picture, Sears has made several trips to Iceland and the Azores to assemble a catalogue of blue whales in the eastern North Atlantic that complements his work in Canada. By 2007, it included well over 100 individuals off Iceland and about 40 from the Azores. Based on what he already knew about the wide-ranging habits of blue whales, Sears was fairly sure that some of the animals he photographed and identified off Iceland would also appear near the Azores. He estimated they would see a match as soon as they had about 30 individuals off the Portuguese islands, and sure enough, in the spring of 2006, blue whale number 32 was ID'd from the Azores and it turned out to be an animal previously seen off Iceland. The two sightings were separated by some 25 degrees of latitude, or more than 1,600 miles.

What was really surprising, however, was a match that came in 2005, when an Australian researcher photographed three blue whales off the coast of Mauritania, in northwest Africa. Sears recognized a prominent notch in the dorsal fin of one of the animals, and was amazed to find it matched a whale seen off western Iceland in 1997 and 1999. He had been expecting to find the Iceland whales off the Azores, but matching one as far south as the Sahara Desert means that some of these animals are making enormous north–south migrations. "So we know there are blue whales off Mauritania, and we know that they migrate to Iceland and further north. Now we need to find out how far south they get." Sears believes the animals may go as far as Guinea, only 9 or 10 degrees north of the equator. Unfortunately, there is virtually no historical data on North Atlantic blue whales at latitudes this low; some are known to have been taken at Cabo Blanco, on the border between Mauritania and Western Sahara, and scientists have twice reported observing blues off Cape Verde. Beyond that, it's all unknowns.

As Sears slowly begins to understand the range and distribution of blue whales on both sides of the North Atlantic, one question continues to come up. Do the whales move east to west and interbreed with one another? Is it possible that Sears might one day get a photo of a blue whale off northwest Africa and match it with one in his St. Lawrence catalogue? That's not likely, he says, noting that if there was regular interaction, he probably would have found a match by now. There is some historical evidence for blue whales crossing the North Atlantic: whalers off Iceland and Finnmark in the 19th century reported hauling in blue whales whose bodies contained broken pieces of American harpoons. (Some historians believe this is simply evidence that American whalers were active in the eastern North Atlantic.) "If right whales can swim from the

Gulf of Maine to the Norwegian fjords, it's very probable that a blue whale could do it as well," Sears says. "But it seems they choose to stay in areas they know, probably where their mothers take them, as is the case with other species. If there's mixing, it's probably not that common."

———————

Although blue whales in the St. Lawrence are primarily attracted to the abundant krill, MICS researchers have long wondered whether there were other attractions as well. "As recently as 10 years ago, most people saw no significant social behaviours on the feeding grounds," says Thomas Doniol-Valcroze, a French-born scientist who joined MICS as an intern in 1995 when he was just 19 years old. "All the socially relevant behaviours were believed to happen on the breeding grounds. Humpbacks and finbacks will feed in groups, but since blue whales are solitary animals, it was assumed that they were just showing up at the same place because there was a lot of food there. People had this very clear-cut view of the breeding season and the feeding season."

Blues do indeed appear to be loners, at least from observations at the surface. Among the thousands of encounters that MICS has recorded in the St. Lawrence, about 83 percent has been animals feeding or travelling alone. Only 1 percent has included three animals, and Sears can count the number of "quatros" (groups of four) on one hand. Of course, all animals have to come together eventually if they want to propagate the species, and pairs of blue whales can sometimes be seen swimming together for several hours at a time during the feeding season. MICS researchers define a pair as two animals seen within one body length of each other for at least one surfacing, and who appear to be synchronizing their breathing, diving and direction of travel. Beyond this coordination, however, there is no conspicuously social behaviour. "The animals pop up together when they surface and then that's it," says Doniol-Valcroze.

"We don't even know what they are doing between surfacings. They might be swimming side by side under water, but they might also be going their separate ways and then coming back together at the surface. We have absolutely no clue."

Richard Sears has long been interested in what goes on between these pairs. If the couple is made up of a male and a female, why are they together during the summer, when it seems that courtship and mating take place in winter? If the whales are the same sex, does that suggest they are feeding together, perhaps cooperatively like humpbacks are known to do? (Humpbacks often work together to herd fish by swimming around them in gradually shrinking circles.) Interesting questions, but none of them could be answered until scientists knew the gender makeup of the pairs.

It is almost impossible to determine the sex of a blue whale in the field. They rarely expose their undersides at the surface, and even if you get a glimpse, the genital areas of males and females are outwardly similar. Sex determination remained out of reach until scientists developed a technique for accomplishing it by genetic analysis. All they needed was a sample of tissue that contained DNA. In most cases, researchers obtain these by driving a boat alongside a surfacing whale and using a crossbow to fire a bolt tipped with a stainless-steel tube. Behind the tube is a foam collar that prevents the tip from going in more than an inch or so before it bounces off with the sample. The scientists usually fire from 20 to 65 feet away, and there is no retrieval line, so finding the arrow in the waves is occasionally something of a challenge. This is rarely a problem if Sears is driving the boat, however. Doniol-Valcroze remembers a day when he dropped his only pencil out of the boat as the team was racing after a whale they wanted to photograph. Sears caught up to the whale, got his photo and then promptly circled back and found the tiny pencil bobbing on the waves. For those with lesser navigational skills, the biopsy arrows are brightly coloured and usually easy to spot.

The tissue removed during these biopsies includes a thin layer of skin and a larger plug of blubber. The skin is sliced off and sent to a lab, where the DNA is analyzed to determine the whale's sex, while the blubber is often retained to be analyzed for contaminants. Obviously, for gender information to be useful the researchers must also photo-identify the whales from which they gather their samples; over the years Sears has learned the sex of about 40 percent of his blue whale catalogue. Just as photo-ID allowed researchers to monitor the movements of individual blue whales across the years, the newly acquired ability to determine the sex of the animals opened another window on their lives. Now it was possible to determine the gender makeup of the blue whale pairs and trios on the feeding grounds. They could even apply this new tool retroactively—during their years in the field, MICS researchers had logged many observations of how individual whales were behaving in pairs and small groups, even though they did not know the sex of the animals at the time. Now they could biopsy those individuals and look at their past behaviour in a new light.

To analyze this new information, Sears enlisted the help of Doniol-Valcroze and Christian Ramp, two of his more mathematically inclined colleagues. "When Richard asked us to begin working on the data, he already had his gut feeling about what the answers were going to be," says Doniol-Valcroze, "but he needed the right statistical framework to make a rigorous analysis." Sears felt there was indeed socialization happening on the feeding grounds, and he believed the blue whale pairs would prove to be male–female, not animals of the same sex. "At first we found results that were along the lines of what Richard was predicting, but there was some noise. There were a lot of pairs that were *not* made up of one female and one male like he was guessing they would be." Now the challenge was figuring out whether any patterns in the data might predict which pairs were likely to be of opposite sex. The key factor turned out to be the length of time the pairs stayed together: the longer they took to separate, the more likely they were to be one male and one female. All the scientists had to do now was draw a statistically valid line between what they had termed short-term and long-term pairs. "We didn't just want to find a limit that would agree with what Richard was thinking. We had to find a threshold that seemed to make sense from an ecological point of view, and that clearly separated the pairs into two categories. In the end, after looking at the data, we decided that the limit was one hour. And as soon as we did that, the data became clear. It really seemed to separate what was a fleeting encounter on one side and a more meaningful encounter on the other."

If one hour seems like a long time for a fleeting encounter, remember that blue whales can dive for 20 to 30 minutes at a time. The short-term pairs, then, were animals that came within one body length for just one or two surfacings. Of the 48 short-term pairs for which the MICS team knew the sex of both animals, 25 were mixed, 13 were both female and 10 both male. In other words, the distribution was entirely random. But when they looked at the long-term pairs for which they knew both sexes—whether the animals stayed together for 60 minutes or several weeks—the results were anything but random: 62 of the 68 pairs were male–female. The analysis also showed that blue whales form more of these long-term pairs toward the end of the summer. This makes logical sense; when they first arrive in a feeding area they concentrate on filling their stomachs, but as fall approaches they spend more time finding a partner for the upcoming mating season. More than 84 percent of the long-term pairs were also made up of animals who were at least seven years old, meaning that these probably were not simply playful adolescents, but sexually mature adults. There may not be any full-fledged courtship going on, but clearly there is some socialization between male and female blue whales on the feeding grounds.

Sears had a gut feeling about something else. He had long noticed that blue whale pairs tended to align themselves so that one animal leads and the other follows, not directly behind, but off to the left or right. The animal in front also tended to be the larger of the two, so he predicted that the mixed pairs were primarily a female leading a male. When the MICS team examined their data, Sears's hunch was dramatically confirmed: for the pairs in which they knew both sexes, it was *always* the female in front. "The first time we put the figures together with the data we had, it was a hundred percent," says Doniol-Valcroze. "Richard said, 'I can't present this as a hundred percent. People will think I fudged the data.'" The trend was so strong that when they looked at long-term pairs for which they knew the sex of just one animal, they could be almost certain of the other even without a biopsy. If the known animal was male, it would be the flanking whale. If they knew that one animal was female, it was almost always the lead one. "Of course, there's more noise here, because if it was a same-sex pair you would find females flanking other females or males leading other males. But the data are still very clear-cut." Sears feels the most likely explanation is that by lagging behind and staying off to the side, the male can keep track of his potential mate and ward off others who may try to approach her.

Armed with this new finding, Sears and his team set out to see what they could learn about trios. Blues can appear jittery when they are approached by boats, but they are not usually given to the dramatic behaviour he often sees in humpbacks, who will slap their pectoral fins and tails on the surface and trumpet loudly through their blowholes. Over the years, however, Sears learned to expect excitement when three blue whales appear together. He would be watching a pair when suddenly a third animal would pop up close by, usually for a single breath, and then dive again quickly. The next thing he knew, all three whales would surface together forcefully and the whales would begin coursing—that is, racing just under the surface—and when they came up to breathe they had so much momentum that they could breach. Sometimes one or more of them lets out a loud, rumbling exhalation that seems to indicate agitation. Sears likens it to the violent stutter of a truck hitting its air brakes as it rolls down a hill. These aggressive interactions, which Sears calls "rumbas," happened in about 80 percent of the trios the MICS researchers observed, and sometimes got violent. In September 2007, Sears witnessed one of the most spectacular encounters of his career when he happened upon Pewter (B185) flanking a whale the researcher recognized as a female. When a second male approached, Pewter and the interloper began "beating the tar out of each other," ramming heads and churning up the flat-calm surface with their thrashing tails and fins. While most rumbas last 10 or 12 minutes, this one went on for almost half an hour. At one point, after the whales had temporarily disappeared under the surface, Sears looked behind the boat and suddenly saw "three missiles coming out of the water." He hollered to the driver to get moving, but she froze, and for a second he thought their number might be up. They eventually got the boat moving and the whales splashed down just behind the stern, to the relief of the biologists and the British tourists who were riding along.

What could be riling up these normally placid animals? "The only thing Richard could imagine that would make the animals do this was competition for females. I don't know how much of his personal background he put into that observation," jokes Doniol-Valcroze. It turned out that Sears was right. Whenever they learned the sex of all three animals, it was always a male–female pair being approached by a second male, and it was the two bulls that were having at it. Sears believes the second male is attempting to insert himself between the pair, forcing the first male out of position. He also suggests it might be the females who initiate the coursing behaviour, to see how adept their suitors are at keeping up.

Can all of this be considered blue whale courtship? Not exactly. "You could just say it's familiarization between individuals," Sears explains. He feels the animals may just be trying each another out to see whether there is a spark. If there is, perhaps the male will approach the same female months later with a nudge, a wink and a "Remember me?" There may also be some advantage to pairing on the feeding grounds. "I think the female accepts the male following her because if he's able to keep others away it allows her to feed in relative peace. If a female is harassed by other males, then she spends her time getting away from them and not doing what she needs to do, which is feed. On the other hand, she also needs to mate to pass on her genes, so she has to be somewhat receptive. On the male side, it's up to him to be accepted as long as he can in the hope that, by spending a few weeks with this female, he's the one that eventually gets to mate with her. Is that what happens? I don't know." In order to complete the picture, Sears says, he would need to observe the whales in December or January when the mating is actually taking place. "We're at the bar at 10 or 11 at night, but we're not there at closing time. All the good stuff happens at one in the morning."

While blue whales almost certainly do not mate exclusively with the same partner for life, pairs may stay together for several years. One suggestion of this comes from the marking program that the Soviet Union carried out in the 1960s. On November 23, 1962, Soviet whalers tagged several blues in the Antarctic as part of their research on the animals' movements. Another expedition, operating more than 13,000 miles away, recovered two of these tags on January 2, 1967. One of the tagged whales killed was male, the other female. It might have been a coincidence. Then again, perhaps these blue whales were a breeding pair that had travelled together over countless thousands of miles for more than 50 months—till death did them part.

SIX

WEST COAST BLUES

ON A JULY MORNING IN 2005, John Calambokidis backs his red Toyota T100 down the boat launch at Ventura Harbor in southern California. With the practised ease of someone who has done it hundreds of times before, the biologist slides the white and grey rigid-hulled inflatable from his trailer into the water. He pulls on a nylon hat with flaps to shield his ears and nape from the sun. Then, before donning his orange survival suit, he slips a personal emergency beacon around his waist. It's a sensible precaution for a scientist who regularly takes a 20-foot boat up to 50 miles offshore, often working solo. "There's an inherent risk in being that far offshore in a boat that size," he admits, "and I recognize that it's not the best way to operate. But sometimes when I'm travelling I don't have anyone with me, and there's also a certain enjoyment in being out there alone."

In addition to equipping his boat with twin engines and carrying backups of the key electronics, Calambokidis carries more than two decades of experience in these Pacific waters. For a couple of months each year, he lives a life he describes as "boat by day, drive by night," during which he and his colleagues use three inflatables to cover most of the coastal waters between British Columbia and Mexico. "I think I've launched out of just about every harbour between L.A. and Washington," he says. Not only has he never needed to be plucked from the sea, he's played the role of rescuer himself on several occasions, towing in a 40-foot sailboat, aiding two teenagers in a capsized Hobie Cat and carrying in a soggy sailboarder. Nonetheless, ships that spot Calambokidis alone on the waves sometimes come by and offer assistance. "The common line we used to get from the Coast Guard was, 'Where is your mother ship?'" The weirdest rescue attempt came one day while he was working about 30 miles off Oregon, when a fishing boat pulled alongside and asked if he was in trouble. The fisherman said he was alarmed by the flares he had seen, a comment Calambokidis couldn't understand. "It wasn't until later, after talking to another researcher who had also encountered this guy, that I figured it out. He must have seen the blows of the whales I was with and mistaken them for flares."

Not that there haven't been frightening moments in heavy seas; his boat has been swamped several times, though rigid-hulled inflatables are relatively easy to bail and tough to sink. The Santa Barbara Channel is also a busy shipping lane, and on foggy days a cargo ship could pick off a lone scientist like a mosquito on the windshield of a tractor trailer. "There have been two or three times where I've had to take very rapid evasive manoeuvres when a ship came into view less than a hundred yards away, bearing down on me." Then there are the whales themselves, which can sometimes get skittish when you race alongside them and try to get a biopsy with a crossbow. Even the friendly ones can be dangerous if they decide to use your boat as a plaything. One summer, while Calambokidis was alone in the channel, a pair of humpbacks approached his boat and one deliberately gave it a shove with its huge pectoral fin. Another actually lifted the boat with its back—again, he is certain this was intentional—taking it well out of the water and then placing it down again gently. "I can't think of a time where I thought, 'Gee, I'm not going to make it here,' but there are lots of times when you get scared on the water. I think that's a healthy thing. It has to do with respecting the dangers that are out there."

Calambokidis is tall and wiry, with a greying beard that he strokes thoughtfully when he's considering a question. Born in 1954, he graduated from Evergreen State College in Olympia, Washington, and in 1979 cofounded the nonprofit Cascadia Research Collective in that city, where he lives with his wife, marine biologist Gretchen Steiger, and their children, Alexei and Zoe. Cascadia now employs about a dozen researchers, and its annual budget of about US$1 million is funded by federal and state agencies and environmental foundations. For almost 30 years, Calambokidis has been studying whales and other marine mammals all along the west coast of North and Central America. Like Richard Sears, he never got a graduate degree, though he does serve as adjunct faculty at Evergreen and he is diligent about getting his research published in peer-reviewed journals. Also like Sears, Calambokidis's physical endurance is legendary. On this day in Ventura, he will spend more than five and a half hours on the water in the July sun without once taking a bite of food or a sip of water. After he returns to shore, he will immediately deliver two public presentations that take another three hours, still without so much as a granola bar.

Calambokidis is today one of the world's leading researchers on blue whales, but by the mid-1980s he hadn't even seen one. "I had been studying marine mammals for seven or eight years at that time," he says, "but I had never encountered a blue whale, and wasn't sure I ever would. Many scientists in the 1960s and 1970s thought that maybe our hunting of the species had driven it to the brink of extinction, and that future generations wouldn't be so fortunate as to see blue whales." In 1986, however, he was involved in a survey of humpbacks in the Gulf of the Farallones, off the coast of San Francisco, and he got a surprise visit. "I knew right away what it was, not just from the size, but because it really was *blue*. I thought, 'What an unusual sight—I can't believe we've actually seen one.' And then we saw another, and another, and another. In that initial year, we estimated there were a hundred blue whales in the Gulf of the Farallones Marine Sanctuary."

Over the next two seasons, Calambokidis surveyed the waters off central California from ships and from above with a Cessna 172 aircraft. Each year he and his colleagues saw more of them. They photo-identified 35 individual blue whales in 1986, followed by 75 the next year, then 101 in 1988. "At that time, no one had any idea that many blue whales existed off our coast," he says. After these findings were published in 1990, Jay Barlow of the Southwest Fisheries Science Center in La Jolla, California, conducted a thorough survey to estimate just how many blues might be out there. In a 175-foot vessel called *McArthur*, he covered an area of more than 300,000 square miles and ventured more than 340 miles offshore. At the end of a journey that took more than three months, Barlow estimated there were 2,250 blues off the California coast, making this population the largest in the northern hemisphere.

Scientists studying whales use two different techniques to estimate population size. Barlow used what is called the line-transect method: the ship travelled in a grid pattern, like a search party, while two observers perched above the deck scanned the sea with high-power binoculars and a third kept lookout with unaided eyes, recording the data on a laptop computer. Whenever the team spotted a marine mammal, the ship would alter course to get closer, and the observers would try to identify the species and, if there was more than one animal, the size of the group. In this way, the researchers sighted 127 large whales, of which 49 were blues. At the end of the survey, they ran the data through an intimidating mathematical formula that includes the number of sightings, the average group size, the size of the study area, the probability of making a sighting along the track line and other factors. When the data are properly processed, the line-transect method is a reliable measure of the density of the population within the study area. Barlow's team conducted several such surveys in the 1990s, and by the end of the decade they were reporting about 3,000 blue whales off California.

The second method for estimating population size is called mark-recapture. This is different from the line-transect method in two important ways. First, each whale must be uniquely identified, not just counted. And second, the researchers must do two surveys, separated by at least a year to allow for normal mixing, mortality, and for animals to leave or join the group. The idea is that you collect two random samples, determine how many animals appear in both, and then use these figures to estimate the total size of the population. Here is a simplified example: imagine you want to estimate how many marbles are in a box. Using the mark-recapture method, you would randomly pull out any marble, mark it with an X and then replace it. You'd repeat this a significant number of times—60 would do the trick—shaking up the box after each marble is replaced to ensure true randomness. Then you would pick out another batch of marbles in the same way—this time let's assume you get 50. Of the marbles in this second group, you notice that eight are marked with an X—these are the "recaptured" marbles, or the ones that occur in both samples. Multiply the number of marbles in each sample (60 x 50 = 3,000) and divide this by the number of recaptures (3,000 ÷ 8) and you get the approximate number of marbles in the box: 375. (Note that in this example, there is no mixing with marbles from other boxes during the period between the two samplings. In a population of whales, by contrast, some will die, be born, or leave or enter the study area during the interval.)

It is important to point out that the two techniques do not just vary in methodology; they measure different things. Line transects count the number of whales in an area at a given time, while mark-recaptures estimate the overall size of the population that uses that area during some part of its life. If all the animals in a population are confined to the study area, like marbles in a box, this distinction becomes meaningless. But blue whales move between many different areas with no predictable pattern. To understand why this is significant, imagine a group of 2,000 whales that travels regularly between points A, B, C and D, each of which is 1,000 miles from any other point. Your study area is located somewhere amid these four points, and at any given moment, only 500 will be within its boundaries. The line-transect method would give you an estimate of 500, while mark-recapture would yield a result of 2,000.

So while Barlow's team counted the blue whales in the waters off the western United States, Calambokidis used mark-recapture to estimate the overall size of the population that spent at least some time there. "Marking" the whales in this case simply meant getting a clear photograph that could be compared with those already in his database. Between 1986 and 1997, Calambokidis had compiled a catalogue of just over 1,000 individual blue whales. (By 2007, that number had almost doubled.) Because he knows the location and date of each photograph, he can look at all the photo-IDs from one year and compare them with those made a year or two later and see which whales were photographed at both times. Using this method in the late 1990s, Calambokidis came up with an estimate of about 2,000 blue whales for the eastern North Pacific population. Given the considerable margin of error, the two methods were largely in agreement.

The discovery that blue whales off California were more numerous than anyone had imagined presented an enormous opportunity for Calambokidis. Here was a chance to work among the largest and most majestic of the ocean's animals, while staying close enough to shore to be able to get back for a nice supper and a good night's sleep (although Calambokidis has been known to sleep in his truck). And the research would be almost entirely new; the most basic questions about this population had never been asked, let alone answered. "What motivates me as a scientist is the excitement of learning something new, something no one else knows, and then sharing that with the world," Calambokidis says. "That's what excites me about working with blue whales—there were so many mysteries, and many of them are now getting resolved."

As he leaves Ventura Harbor this July morning, Calambokidis opens up the 115-horsepower outboard and is soon doing 25 knots, bouncing gingerly over the small swells. He's as sure-footed as a mountain goat, casually steering with one hand as he takes a call on his cell phone, as though he were cruising the Pacific Coast Highway in his Toyota. As the shore recedes, it becomes clear that studying blue whales off California is quite different from working in the St. Lawrence estuary: you're often well out of sight of land. It's a good three-quarters of an hour before he slows down and begins his search for blues. He starts by checking his echo sounder, an instrument that sends out two sound pulses at frequencies of 50 and 200 kHz. The lower pulse travels deeper and farther, he explains, but it tends to detect only large objects, such as fish. The higher frequency, while it doesn't detect faraway objects, is better at spotting smaller creatures, including krill. By comparing the intensity of the readings at the two frequencies, he can determine if there is a krill patch in the area, and if so, at what depth. Where there's krill, there are often blue whales.

Around 1 p.m., Calambokidis spots a blow almost a mile away. He turns the boat sharply to the right and motors off in pursuit. When he catches up to it, the whale surfaces again with a breath like a loud sneeze. As the whale rolls through the water before diving, several seconds pass between the time the blowholes are visible and the appearance of the tiny dorsal fin, a testament to the enormous length of the animal. It is the markings on the back of the whale, near the dorsal fin, that Calambokidis wants to capture with his camera, and he snaps a few frames of the whale's left side, which is unusually light-coloured. Over the next several minutes, the whale surfaces a few more times before diving out of sight. With experience, it's often easy to predict when a whale has completed its so-called surface series, a sequence of breaths followed by a deep terminal dive, after which the whale disappears for at least five or ten minutes, sometimes much longer. "They change their angle on their last dive and go almost straight down," Calambokidis explains. "You'll see a slight increase in speed right before they make their last surfacing, and they'll arch high and potentially show their flukes. I think that's to get some forward momentum, and then they use their pectoral fins to change their body angle downward."

Once the whale dives, it is difficult to predict where it will come up next. "It's a pretty imperfect science," Calambokidis says of anticipating a whale's movements. "If you're with a whale for a while you get to see what it's doing, but if you've only been there for one surface series then you have to make a decision about whether this is a travelling whale or a feeding whale. Should I stay in the same place, or should I move along at 3 or 4 knots in the direction I think it's going? Part of that you can predict by what other whales in the area are doing and what the prey layer is like. If you've got a dense krill layer and other whales are feeding in the area, then it's a good bet the whale is going to stay."

It also helps to have keen senses. Calambokidis, Sears and other experienced researchers can spot a whale blowing a mile or two in the distance even on hazy days, and they can hear an exhalation over the sputtering of an outboard motor. Sure enough, some 20 minutes after his first encounter, Calambokidis perks up his ears. He turns the boat around, tears off in the direction of the sound and quickly finds his whale again. He is pretty sure it's the same one and he is determined to get a photo of its right side this time, so he'll have a more complete record for his catalogue. He does just that on the next surfacing, and then pulls out a battered black clipboard to make his field notes. After each encounter with a whale, Calambokidis records the date, time, GPS coordinates, water depth, the size, shape and location of any prey patch showing up on the sounder, and even any birds that may be feeding nearby—krill-eating birds such as Cassin's auklets can help identify the prey species the whales are feeding on. He uses a series of codes to describe specific behaviours when he sees them: a surface lunge, a breach, an attempt to circle or avoid the boat. If he sees a pair or trio, he will note which animal is leading and which is trailing, whether it is a mother and calf, and any size or behavioural difference between the animals. Later that afternoon, Calambokidis does encounter a trio, and he moves to the front of the boat, removes the protective cover from his crossbow and slides in a biopsy dart. He gets close enough to biopsy one of the three whales, then plucks the crossbow bolt from the sea, removes the sample with tweezers and places it in a red Thermos bottle. When he gets back to shore, he'll freeze

Biologist John Calambokidis records his observations in the field. Calambokidis has been studying blue whales in the eastern North Pacific since 1986.

it in liquid nitrogen and send it to a lab that will determine the whale's sex.

While Sears is on a first-name basis with many of the blue whales in his study area, the much larger California population makes this impossible for Calambokidis. However, there are a number of distinctive-looking animals in these waters that he recognizes on sight—or on sound, as the case may be. As he approaches a lone blue whose spout he has seen in the distance, the animal surfaces and makes an unusual snort as it exhales. Calambokidis immediately sees that the whale has a large wound near its blowhole that seems to affect its breathing. He has seen this animal before, and though he's not sure what might have caused the wound, he suspects it does not pose any threat. In fact, there are at least 20 other blue whales in his catalogue with significant wounds, and hundreds more that have at least some damage to their dorsal fin.

Toward the end of the afternoon, Calambokidis is sharing his study area with a couple of whale-watching boats. He says it's always a challenge to try to do good science around these sightseers. "I don't want to make close approaches to whales around their boats because I don't want people on board to be pressuring their captains to get closer." Because blue whales are an endangered species protected by federal law, it is an offence to harass the animals—the general guideline is to back off a hundred yards. While most cruise captains respect the rules, they are also keenly aware that they are competing with others who may not be so enlightened. Calambokidis also tries to avoid taking out his crossbow in sight of whale-watching boats, since an orange-suited guy firing darts at whales might be misunderstood. (If it sounds unlikely that a scientist might be mistaken for a hunter, remember the eagle-eyed fisherman who thought whale blows were flares.)

The relationship between scientists and whale-watchers isn't always adversarial. Calambokidis has a good relationship with one boat called the *Condor Express*, captained by his friend Fred Benko. They've found that the trick to healthy collaboration is a balance between sharing information and staying out of each other's way. Benko's boat, one of the best-known whale-watching vessels in the Santa Barbara Channel, goes out every day of the season to cover the same small area, so Calambokidis can learn a lot from Benko about what's going on in those waters. In exchange, the scientist shares his expertise with Benko's clients, who are always full of questions about the whales they're seeing. After Calambokidis has finished his work with the whales he encounters, he'll sometimes even call Benko on the radio and let him know where they are.

The real test of the relationship comes when both scientist and whale-watcher happen upon whales at the same time. Calambokidis recalls one day when he was putting bioacoustic tags on blue whales in the channel. He had found a pair of blues and was just attempting to deploy a tag when he saw the *Condor Express* heading his way. The boat had been out all day and hadn't seen a single whale, so Calambokidis knew its crew was anxious. But he decided to get on the radio and ask them to hold off for a half hour until he got his tag on. "I would rarely do that, because it's a huge thing to ask," Calambokidis says, well aware that these might have been the only blue whales in the immediate area. But Benko was gracious; the *Condor*

Express agreed to hang back, Calambokidis got his tag deployed, and the whale-watchers approached the animals later. On that day, a little diplomacy allowed everyone to share the whales.

———————

Soon after John Calambokidis encountered his first blue whale off California in 1986, he began to ask an obvious question: why had so few blue whales been seen in this area before now? The riddle became even more baffling once Calambokidis learned that this population numbered more than 2,000 animals. Indeed, blue whale sightings off California, not only in the Gulf of the Farallones, but also in Monterey Bay and the Santa Barbara Channel, increased dramatically in the mid-1980s and the early 1990s. Calambokidis wanted to know why.

The plight of blue whales in the eastern North Pacific during the whaling era was much the same as in other regions, which is to say that whenever humans could find them, they did their best to kill them. Charles Scammon described several thwarted attempts to hunt "sulphurbottoms" off Mexico in the 1850s using the newly developed bomb-lance. By 1862, whalers managed to successfully take a blue whale in Monterey Bay that reportedly measured 92 feet. It became grotesquely swollen by the time they had hauled it to shore, "but on arrival, no effort was spared to strip the colossal prize of its fatty covering."[1]

It would not be until floating factories arrived just before the First World War that blue whales here would die in significant numbers. Between 1910 and 1965, whalers took at least 9,500 blue whales in the North Pacific, possibly extirpating them from areas where they were never numerous, such as southern Japan. The vast majority of those killed off North America were taken on their summer feeding grounds, which ranged from British Columbia into the Gulf of Alaska. While the waters off Mexico and California remained popular with whalers, blues made up only a small fraction of the catches there. From 1958 to 1965, several shore stations were operating in California, taking 2,176 whales, but only 48 of these were blues. Most people believed the few blues they did see were simply passing by on their route to the North. No one believed that blue whales spent any significant time in California waters.

The seasonal travels of blue whales have never been obvious, or even consistent, and the animals of the North Pacific are no exception. In 1962, the International Whaling Commission appointed a group of researchers from the U.S., Canada, Japan and the Soviet Union to learn more about the mysterious movements of the region's blue whales. One of those scientists was Dale Rice, an American who did some of the most important early research on the species. Rice and his colleagues did useful surveys in 1965 and 1966, and even marked a number of blue whales using the same method the Discovery Committee pioneered decades before. From the bow of a recommissioned whale catcher, Rice used a shotgun to mark 76 blue whales between the southern tip of Baja and San Francisco, but not a single one was retrieved before the hunting of blue whales was stopped. Soviet researchers did manage to recover at least 15 of their own tags, however, and some of the whales showed remarkable movements. One was marked off Vancouver Island in May 1958 and killed 13 months later off Alaska's Kodiak Island, about 1,110 miles to the northwest. They deployed another steel tag off the southwestern tip of Russia's Kamchatka Peninsula and pulled it out of a dead blue whale 49 months later in the Gulf of Alaska. That's an east–west movement of more than 2,300 miles. These two tag recoveries raised the possibility that blue whales from both the western and eastern North Pacific were visiting the waters off Alaska during the whaling era. And because both tags were recovered in the month of June, perhaps whales from both populations were occasionally in the same place at the same time. That suggestion would take almost 40 years to confirm.

In his 1966 report to the IWC, Rice confessed he was unable to see a clear pattern in the seasonal travels of the animals. "The occurrence of fairly large concentrations of blue whales off Baja California from February to July is difficult to reconcile with the postulated movements of blue whales in the North Pacific Ocean. The season of occurrence does not coincide with the concept of a 'wintering' ground."[2] Why, he wondered, were some blue whales staying in Mexican waters as late as July, when conventional thinking would have them moving north in early spring? Writing again about these surveys in 1974, Rice was forced to make his best guess: "I postulate that the majority of this stock of blue whales leaves Baja California waters in May and heads north, passes California far offshore, and arrives on the whaling grounds off Vancouver Island in June. From there at least some must proceed to the eastern Aleutians or into the Gulf of Alaska."[3] Not all the animals followed this route, clearly, but it seemed to be a general trend. All of the blue whales along the west coast of North America, therefore, probably belonged to a single migrating population, most of whom spent the summer off British Columbia and Alaska and the winter off Mexico. The whales spent several weeks each year moving between these areas, but they did not tarry long off California.

Indeed, until the 1980s, there were precious few blue whale sightings *anywhere* along the west coast of the United States and Canada. They showed up regularly in Mexican waters, but throughout the 1970s scientists in the Gulf of the Farallones had seen only one, and precisely zero had been spotted in the Gulf of Alaska since 1974, despite several attempts to find them. For this reason, some scientists believed that the eastern North Pacific blue whales had historically comprised not one, but two populations: a southern group concentrated off Mexico and southern California, and a northern group off British Columbia and Alaska, the latter of which had been wiped out perhaps by whaling. Things became more interesting in the early 1980s, when scientists working off central California began to report a few sightings each year, and by the time Calambokidis arrived in 1986, blue whales were downright common between July and November. So where did all these whales come from? If they used to bypass California en route to their feeding areas farther north, why were they now stopping for so long? Was it possible that this newly discovered group was related to the blue whales once found in the Gulf of Alaska and off the eastern Aleutian Islands?

Beginning in the early 1990s, Jim Gilpatrick and Wayne Perryman, two researchers with the Southwest Fisheries Science Center in La Jolla, set out to see what they could learn by looking at differences in the body lengths of the California blues and those that had lived farther north. To measure the whales in the Santa Barbara Channel, they used a technique called aerial photogrammetry, which involves photographing the animals from a small aircraft and then, using techniques borrowed from surveying, precisely measuring their length. When they compared these measurements with historical records from whaling stations in the Gulf of Alaska, Gilpatrick and Perryman found that the northern whales were significantly larger. "Our results suggest that the blues feeding in the [Channel Islands National Marine] Sanctuary are morphologically different from the blues killed in other parts of the North Pacific by whalers," they wrote in 1997. "Our work with these whales has also established that they are a distinct population from other blue whales in the North Pacific."[4] Gilpatrick and Perryman continued their photogrammetric work until 2003, going as far south as the eastern tropical Pacific, and they haven't wavered from their initial interpretation. "It looks like these animals that migrate along North America during the summer are about 2 metres shorter than the ones they took along the Aleutians and in the Gulf of Alaska, and to the west, too, off Kamchatka and that area," Gilpatrick says. "These whales are a different population."

When Calambokidis heard these results, a few things didn't sit right with him. He had always been skeptical about the size differences that Gilpatrick and Perryman had noted. Calambokidis and two Cascadia colleagues had done their own aerial surveys in the late 1980s and were aware that the California blue whales were smaller than those reported in the whaling data, but they put this down to methodology: comparing the lengths of live whales with dead ones raises some problems. Whalers measured carcasses in various ways, and each produced different results. The accepted method was to measure in a straight line from the tip of the rostrum to the notch in the tail flukes. Whalers, however, sometimes used techniques that could add an extra yard or two, such as measuring from the protruding lower jaw, including the tips of the tail, or running the tape along the curve of the whale's back. "They may also have wanted to stretch the size of whales they were taking to get above minimum limits, and because there were often incentives for the size of whale being caught," Calambokidis says. "So that line of evidence always seemed a little tenuous to me." (For his part, Gilpatrick says they did exclude some data from Japanese whalers that almost certainly exaggerated the size of their catches.)

There was also the question of why blue whales off California had suddenly become so common. At first he thought the whales may always have been there, and that the whalers were just unaware of them. But the more he thought about it, the less sense that made. "There were whalers going out killing humpbacks, so there seems to be no reason why they would have missed the blue whales. They hunted them when they found them, yet they took very few." Was it possible that the population was finally recovering from whaling after 20 years of protection? Even now, Calambokidis doesn't think so. "Our data do not show any increasing trend, at least for the little-over-10-year period that we've been making these estimates. So even though we have this large and very healthy population, at this point we're not sure that it's growing—it may be a stable population. There are still more blue whales than anyone thought existed in the North Pacific, so it's good news, but we have to be cautious."

The only remaining explanation was that the California animals must have come from somewhere else. As he and other researchers began compiling more and more data, most of it supported this idea. "I'm increasingly seeing evidence that there has been some shift in distribution," Calambokidis says, "because these whales all look like they're part of the same population." He suggests there may never have been a discrete northern population that was eliminated by whaling; the whales once found off B.C., Alaska and the

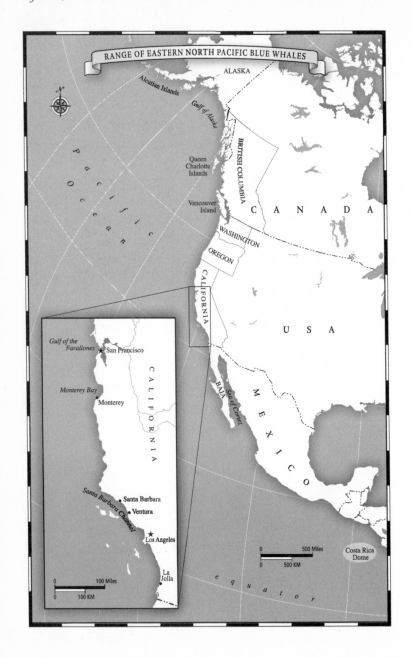

RANGE OF EASTERN NORTH PACIFIC BLUE WHALES

Aleutians were simply part of the wide-ranging eastern North Pacific population that is still alive and well. He believes that some time before the mid-1980s, most of these whales simply stopped moving so far north during their annual summer migration. Instead of heading into British Columbian and Alaskan waters, most simply stopped off California. "Some of them still occasionally go farther north, but because their prey is more abundant off California now, they may be spending more time there."

The first evidence linking the California whales with northern waters came in 1997, when Calambokidis matched a photo of a blue whale taken near the Queen Charlotte Islands, off British Columbia, with one in his California catalogue, clear proof that at least some of the whales feeding off California venture north of the U.S.–Canadian border. Surprisingly, the whale was off B.C. in June and in the Santa Barbara Channel in July; that means it was moving south during the summer, precisely the opposite direction one would expect if blue whales made traditional migrations. In 2003 and 2006, he found two more matches between British Columbia and California.

Another piece of the puzzle arrived in the summer of 2004, when a research cruise led by Jay Barlow spotted three blue whales in the Gulf of Alaska, the first ones seen there in 30 years. A Cascadia researcher aboard the ship immediately sent the photos to Calambokidis so he could compare them with his catalogue. The result: one was a match with a whale seen off California in both 1995 and 1998. As he sees more evidence like this, Calambokidis has come to believe that the blue whales of the eastern North Pacific are a single, widely roaming group that ranges from Central America all the way to the Gulf of Alaska, a stretch of ocean spanning some 50 degrees of latitude, or more than 3,400 miles. Dale Rice's 1974 idea appears to have been essentially right after all. The only modification is that most whales no longer travel as far north, or as far offshore, as they did in the past.

What would cause a population of whales to change its seasonal movements? Why would most of the blue whales that once spent their summers feeding in the Gulf of Alaska and off the eastern Aleutians decide to linger for months off California instead? The short answer is that they are following the food, and the occurrence of krill is closely tied to changes in climate. For example, fish such as capelin, a staple in the diet of many marine mammals, declined drastically in the Arctic between the mid-1970s and the late 1980s, at least in part because of a trend toward warmer temperatures. Over that same period, the populations of harbour seals, northern sea lions and fur seals in the Gulf of Alaska also plummeted. If the number of krill in these waters took a similar nosedive, it would follow that blue whales would stop going there to feed, especially if California waters saw an increase in their favourite prey. The timing certainly fits: these changes occurred during the same period when blue whales began showing up farther south in the summer.

The southern part of their range raised a different set of questions. Early researchers had noted that blue whales were absent from Mexican waters between November and January and speculated that the animals had migrated even farther south. But where? To find out, Calambokidis teamed up in the mid-1990s with Bruce Mate of Oregon State University to employ a new technique: satellite tagging. More about how these tags work in the next chapter, but for now suffice it to say that the method involves attaching a radio transmitter to a whale and tracking its changing position over days, weeks and months. Mate and Calambokidis successfully deployed the tags on 10 blue whales off southern California in the late summer and early fall of 1994 and 1995. On average, the tags in this study stayed on for 29 days (the stickiest one held for 78 days), which wasn't enough to get a full understanding of the whales' movements across more than one season, but nonetheless provided some real breakthroughs.

To begin with, it was immediately clear that blue whales do not all follow the same route when they leave California in the fall. Two whales tagged near one of the Channel Islands, for example, stayed for two weeks before one headed southwest and the other steamed more than 400 miles north. In another case, four tagged whales made what seemed to be the expected southerly migration, arriving in Vizcaino Bay at the midpoint of the Baja Peninsula within two weeks of one another. However, two whales took an inshore route, while the others swung way out to the west and traced a different path to the same location. Most interesting of all, however, was that one of the four continued all the way down the Mexican coast, sending back its last reading 280 miles north of a region known as the Costa Rica Dome. Located about 9 degrees north of the equator, the Dome is an area about 125 miles across, where cool, nutrient-rich water is pushed toward the surface by the convergence of ocean currents. This upwelling results in huge blooms of phytoplankton, the tiny plants that form the basis of marine food chains. The Dome moves from year to year as conditions change, but the upwelling occurs more or less permanently, making it one of the most productive areas in the eastern tropical Pacific, and since phytoplankton gives rise to krill, the region is a logical destination for blue whales. While scientists had suspected for some time that California blue whales may venture that far south, this was the first time anyone had proved it, and it raised the possibility that the Costa Rica Dome might be one of the long-sought wintering grounds of northern hemisphere blue whales. But there's a twist: while the Dome is busiest in winter, blue whales have been detected there year-round. Could this mean that it also supports a resident population that never migrates? Might it be a haven for juveniles who do not accompany the adults on their annual northward journey? Or perhaps the area is visited at opposite times of the year by southern and northern hemisphere blue whales, like Australians and Canadians visiting the same tropical islands to escape their respective winters.

Satellite tagging by Bruce Mate in subsequent years revealed many other blue whales travelling between California and the Costa Rica Dome, and Calambokidis has since made many photo-ID matches between the two areas. Altogether, among more than 1,900 blue whales he has identified in the eastern North Pacific, Calambokidis has matched whales seen off California with more than 90 off Baja, 10 in the Costa Rica Dome, 3 off the Queen Charlotte Islands, and 1 in the Gulf of Alaska. That is pretty compelling evidence to support the idea of a single blue whale population that spans the entire west coast of North and Central America.

What's the big deal about all this? What difference does it make whether the blue whales of the North Pacific are one population, or two, or five? Quite a lot, actually. Bruce Mate explains that we need to know these things if we are going to protect blue whales and help them recover from the slaughter of past decades. "We're trying to figure out what their critical habitats are, and whether these are migration routes, feeding areas or reproductive areas, so we can provide these animals with additional levels of protection where human activities may impede their recovery." He points to the successful comeback of grey whales, which were hunted almost to extinction by the 1930s but have rebounded to the point where they were removed from the United States' endangered species list in 1994. "The only reason those efforts were successful was that we knew their migration routes, and we knew their breeding and calving areas. It isn't adequate just to say 'no hunting.' The Mexican government set aside the grey whales' breeding and calving areas as *refugios*—they don't allow net fishing when the whales are using them. There has to be an understanding of cause and effect."

Since 1991, the International Whaling Commission has formally considered *all* the blue whales of the North Pacific a single "management stock." This includes not only the whales along the west coast of North and Central America, but also those off Kamchatka and the western Aleutians. The World Conservation Union, the organization that draws up the well-known Red List of threatened and endangered species, also lumps them all together. But while Calambokidis and others now believe the animals in the eastern North Pacific make epic north–south journeys, there is no evidence to suggest that these whales spend any time in the western North Pacific. The blue whale distribution map needs to be redrawn.

One of Calambokidis's goals when he founded Cascadia was to provide research that would help protect endangered species. His work with the mysterious blue whales of the eastern North Pacific, however, has humbled him; he is aware that the animals are always several fluke beats ahead of the scientists. Just when it seems as if researchers have some understanding of blue whale behaviour, the animals throw them off the scent. That's why Calambokidis stresses that long-term work needs to continue. "What's happening right now doesn't necessarily mean that's what was happening 50 years ago. And the situation we have today may not be the same in 20 or 30 years. I've moved away from the idea that there's one right answer and that's the way it's always been."

The breath of a surfacing blue whale catches the sunlight and paints the sea air with a rainbow of color. This photo was taken during the 2006–07 IWC–Southern Ocean Whale and Ecosystem Research cruise in the Antarctic. Since the mid-1990s, this annual program has filled in many of the gaps in our understanding of Antarctic and pygmy blue whales. (Isabel Beasley/IWC)

ABOVE: Only about one in seven blue whales raises its flukes before diving. The pigmentation pattern on the tail stock and along the whale's back allows researchers to identify the individual. (Richard Sears) BELOW: The head of an Antarctic blue whale, measured from the tip of the snout to the blowholes, can easily exceed 18 feet in length. (Kate Stafford)

ABOVE: As the sun sets in Quebec's St. Lawrence estuary, Richard Sears has just enough light to attempt a biopsy. DNA from skin samples is the only reliable way to determine a whale's sex. (Dan Bortolotti) BELOW: Flanked by floating bits of ice, a blue whale in the Southern Ocean creates a massive bow wave as it surfaces to breathe. (Isabel Beasley/IWC)

ABOVE: Some of the most animated blue whale behaviour occurs when a second male attempts to insert itself between a male-female pair. (Cascadia Research) INSET: The racing behaviour that occurs among trios can cause the whales to emerge well above the surface. (Richard Sears) BELOW: The only known white blue whale has been spotted in California and Mexico many times since 1999. (Cascadia Research)

ABOVE: A blue whale surfaces languidly in the Sea of Cortez, between Baja California and the Mexican mainland. (Richard Sears) BELOW: While John Calambokidis pilots the boat, a colleague in the bow prepares to deploy a bioacoustic tag on the back of a blue whale, where it will be held in place by suction cups. These instruments collect data on the depth and angle of the whale's dives and record its vocalizations. (Cascadia Research)

ABOVE: The blow of a lone blue whale condenses in the cold Antarctic air. A century ago, some 240,000 blues roamed the Southern Ocean; today, perhaps 2,000 remain. (Kate Stafford) BELOW: A calf surfaces alongside its mother in the waters off Baja. Newborn blues often exceed 20 feet and can gain 40,000 pounds before being weaned. (Richard Sears)

ABOVE: Feeding blues often turn on their side when lunging at the surface. The pleats under the whale's jaw expand and allow the animal to engulf enormous quantities of krill and seawater. (Cascadia Research) BELOW: Although many scientists believe blues can live as long as humans, the oldest known individual is Nubbin, who was photographed in 1970 and again in 2008. (Richard Sears)

When fully expanded, a blue whale's ventral pouch can hold an estimated 13,000 gallons of water, giving the animal the appearance of a 75-foot tadpole. A blue's feeding lunge, which can generate a force that one scientist estimates at 90,000 newtons, has been called "the largest biomechanical action in the animal kingdom." (Richard Sears)

SEVEN

BLUES' CLUES

THE FIELDWORK of Richard Sears in the Atlantic and John Calambokidis in the Pacific contributed as much to the understanding of blue whales as that of any researcher before them. Both have assembled catalogues of the individual whales in their study areas, and both have an overall sense of the animals' seasonal movements, even if the precise itineraries will never be predictable. But both would freely admit that small boats and telephoto lenses cannot solve all the mysteries of the blue whale. How far offshore do they venture when looking for food? How deep do they dive? What feeding strategies are they using underwater? Answering these questions requires technology that can follow the animals when they are too far from a harbour or too deep under the surface to observe directly. By the mid-1990s, this technology finally became available in the form of satellite tags, underwater cameras and bioacoustic probes that can go where no scientist can: right onto the back of a blue whale.

The idea of tagging blue whales—that is, attaching instruments to the animals and later recovering them to obtain data—goes back to the Discovery Committee in the 1930s. But the steel tubes fired into the blubber of whales by these early researchers were nothing more than numbered markers, and the only information they provided about the animal's life was where it was tagged and where it was killed. In the 1950s, scientists took a giant step forward by using radio telemetry to track wild animals: by attaching a collar fitted with a VHF transmitter to a tranquilized animal, researchers could learn its position by picking up the signal on a receiver. This technology was fine for forest mammals, but the limited range of VHF signals, and the fact that they do not travel well through water—not to mention that attaching such a device on a whale would be a formidable challenge—made the technique unsuitable for studying the long-range movements of cetaceans.

In 1978, the U.S. and France collaborated to create the Argos system, a network of satellites and ground stations that can be used to determine the location of any object fitted with a UHF transmitter. This new technology, a predecessor to today's global positioning system, or GPS, enabled wildlife biologists to track migratory animals over great distances. UHF signals still don't carry well in seawater, but as long as a tagged animal—whether it's a seal, turtle or whale—comes to the surface occasionally, the satellites can detect the transmissions and relay the information to receivers on the ground, where it can be used to work out the animal's position. That was the theory, anyway, but it took a long time before anyone designed and built a satellite tag small and robust enough that it could be attached to a marine mammal without restricting its movement. By the late summer of 1994, no one had attempted to use satellite telemetry to track blue whales during their mysterious migrations, and it would take the complementary skills of Bruce Mate and John Calambokidis to make it possible. Mate has been radio-tagging large whales since the late 1970s, beginning his work with grey whales in the lagoons of Baja. As a graduate student he determined the migration patterns of northern sea lions in the Pacific, and in 1989 and 1990 he successfully satellite-tagged right whales and bottlenose dolphins. Mate has since used the technique on fin whales in the Mediterranean, bowheads in the Canadian Arctic and humpbacks around Hawaii. Calambokidis, meanwhile, knew the habits of blue whales off California better than anyone, and he had the boat-piloting skills that would be essential for approaching the animals and deploying the tags.

Mate recognized that satellite tagging could add a new dimension to blue whale research. "I love photo-ID, and I love that it's noninvasive, but it has some limitations, too," he says. "One is that you only get information about where whales go by sending photographers to those places, and you can't afford to do that everywhere and at all times." The other shortcoming of photo-ID is that it encourages researchers to concentrate their efforts in areas that blue whales visit reliably rather than striking out on voyages of discovery. "You don't have the opportunity for big surprises. The first humpback whale we tracked from Hawaii to its summer feeding grounds went to the Kamchatka Peninsula. That was a jaw-dropper for everyone because nobody knew that whales from Russia went to Hawaii. It's these kinds of things that you'll never find out from photo-ID."

Mate and Calambokidis collaborated on their first tagging effort in August 1994 and continued the study the following year. They used two tag models, both about 7 inches long and less than 2 inches in diameter. While one type also included a sensor to measure dive depth, the essential component of a satellite tag is a transmitter that sends out a simple, monotonous signal. Working around the Channel Islands off Santa Barbara, the researchers used a crossbow to fire the tags into the skin and blubber of 10 blue whales, where they were held in place with barbs. Then they waited to see where the whales would go. (Satellite tags have evolved greatly in the past two decades, though none of the designs are wholly benign; to observe the long-term movements of whales with several thousand dollars' worth of hardware, you need to make sure the tag will stay put for months. While the barbs probably cause the whale short-term discomfort, there is no evidence that they do any lasting harm.)

The first thing one needs to understand about satellite telemetry is that it does not work like GPS. The navigator in your car or the hand-held unit you might carry while hiking can pinpoint your location within a few yards by accessing a constellation of 24 satellites, at least six of which are above the horizon at any time everywhere on earth. The Argos system uses just two satellites that circle the earth perpendicular to the equator, passing over the North and South Poles. This means the chances that a satellite is overhead at any given time and place is relatively small. One or the other sweeps above each spot on the planet between 6 and 28 times per day, with that number decreasing as you get closer to the equator, and it takes no more than 16 minutes for the satellite to zoom from one horizon to the other. Add this to the fact that whales spend most of their lives underwater—which renders the tag's signal too weak to reach the satellite—and it's clear that windows of opportunity are few and far between. To get around these problems, the blue whale tags were programmed to transmit only during times when there was a high probability of a satellite being within range (they pass by around the same local time each day). To conserve battery life—a perennial problem with instruments designed for long-term use—the tags were also fitted with a saltwater conductivity switch, which turned off the transmitter whenever the whale was submerged.

Only when everything comes together—a satellite is above the horizon, the tag is switched on and the whale is breathing at the surface—do the scientists get a usable signal from one of these tags. With enough transmissions, they can work out the whale's position by measuring the Doppler shift of the signals. The Doppler effect is the phenomenon that causes an ambulance siren to sound higher when it's speeding toward you and lower as it moves away: the sound waves are either compressed or stretched out by the motion of the vehicle, causing them to reach our ears at different frequencies. In a similar way, each signal that the tag sends during a whale's surfacing is identical, but the fast-moving satellite will hear each tone as slightly higher or lower than its true frequency. By plotting these differences on a curve, the scientists can estimate the timing and the angle of the satellite's closest approach, and thereby determine the latitude and longitude of the tag. Just how accurate this reckoning is depends on the number of messages received during each satellite pass. If the whale happens to surface four or five times when the satellite is visible, the location may be accurate within 500 feet; in other cases, Mate had to allow for an uncertainty of up to 7 miles. That sounds like a lot until you consider that one of the tagged whales in this study travelled 534 miles in just over four days. "[Seven miles] is something a blue whale can do in an hour," Mate says. "We're talking about an animal that is resident to a third of an ocean basin."

Mate's early efforts at satellite-tagging blue whales confirmed that some animals that feed off California in summer do indeed travel much farther south than Mexico, into the vicinity of the Costa Rica Dome. He continues to be amazed at the huge variability between individuals, however. One group of blues that Mate tagged in the Sea of Cortez in March and April 2006 ended up lingering off the west coast of Baja for as long as five months, never getting to California at all. In September of the same year, he tagged a dozen blues off California, and by the end of October they were spread between Alaska and southern Mexico. Why might individuals feeding in the same area head off one day in entirely different directions? What factors influence their decisions to forage in one place as opposed to another? Do they remember favourite feeding areas, or are they somehow able to read environmental cues, such as sea surface temperature, from great distances? "I wish I could be definitive about this—ultimately, that's where we'd like to be," says Mate. "But we have not tagged the same animal twice, so we don't know how much variability there is for an individual. If we knew that the same whale did the same thing year after year, then we could attribute much of what they're doing to past history, whether it's learned from their mother or from experience. Without that, when we see variability between individuals, or from year to year, we just can't ascribe it. But it is clear that in a specific year different individuals have different strategies."

"Even when you have a stable feeding area, individual blue whales are constantly going off somewhere else and then coming back," Calambokidis adds. "It's like they're constantly checking for someplace better, and if they don't find it, then they return. If they *do* find it, they don't come back. I now realize that even when you see blue whales in the same spot, it's not always the same individuals."

Mate's work has also made it clear that there is a lot of individual variation in migration, both in the route the whales take and the timing of their departure. "Eighty-five percent of the grey whale population will sweep along the southbound migration route off California in a period of just three weeks," he says. "It's a very compact thing. But what we've learned from tagging blue whales is that the migration toward the breeding grounds is certainly not a coherent wave, with the bulk of the population moving in a short period of time. Blue whales are making decisions based on their own reproductive state and fitness with a great deal more variability than we see in grey whales."

While the unpredictability of blue whales poses a challenge for scientists, Mate points out that there is an upside. "This was probably a huge saving grace for blue whales. It's prevented them from being exploited to extinction by whalers. The thing that brought grey whales into jeopardy was their very predictable near-shore migration." Grey whales arrive reliably in the Sea of Cortez each spring, where whalers would be waiting for them. "Blue whales, on the other hand, by going to these far-offshore places that change from year to year depending upon oceanographic conditions, would have made it extremely difficult for whalers to find them in the first place, and to return to those areas dependably in subsequent years. Because blue whales were the ultimate target of the industry, I think this situation really saved their bacon."

One of the most tantalizing promises of satellite tagging is that it may provide solid evidence of something researchers have long suspected. "I think that the Costa Rica Dome is going to turn out to be a calving area for blue whales," Mate says. This would be a momentous finding, given that no such area has ever been confirmed in any ocean, at least nothing that resembles the Mexican lagoons used by grey whales, or the several calving grounds visited annually by pregnant humpbacks. Calambokidis, Sears and others regularly see blue whale calves in the Sea of Cortez that are a couple of months old, but they have never seen a newborn. Indeed, aside from a couple of little-known reports from the first half of the last century, no one has observed a newborn blue whale anywhere in the world.

So what leads Mate to believe that the Costa Rica Dome might be a nursery? Some of the blues he has recently tagged seem to show the same characteristics as pregnant females of other species when they go to their reproductive areas. "We've tracked animals going to the Dome early in the winter and staying a long time," he says. "Usually pregnant females are among the leaders in grey whale and humpback migration. They get to the breeding and calving areas first and they stay longer than other animals because they're waiting for the calves to grow up and get strong enough to make the migration away. If some blue whales are going to the Dome and staying in a specific area for months, that's not typical of an animal that is just coming to breed and then going back to the feeding grounds." Knowing for sure, however, would require an extremely costly expedition, and the timing would have to be just right. "It would have to be early enough in the season where you arrive before the calving starts, so you see them as newborns." The trip would also have to take place during a season when food is plentiful, since that would increase the number of whales that are fit enough to reproduce.

"So we're doing another approach that's kind of novel," says Mate. When his team tags a whale, they take a small biopsy at the same time. Working with experts in hormone detection, Mate is trying to determine whether it's possible to tell if a female is ovulating, pregnant or lactating from the tiny amounts of estrogen and progesterone in the sampled tissue. "The problem is that the samples are skin and blubber, and it would be more reliable if we could draw blood and look in the plasma for these concentrations. As it is, we're looking at micro levels of these hormones in fatty tissues, so we're facing a big challenge. But if we can get that to work, we can go back historically and look at all the animals we've tagged and biopsied and ask, 'Do the levels of estrogen and progesterone in the animals that went to the Costa Rica Dome suggest that they are mostly pregnant females?' This is something that's really important for the future. We wouldn't have to mount an expedition if we could interpret this stuff routinely. But it's very hard to find someone to fund the research."

For now, Mate continues to tag about a dozen blue whales annually, and his lifetime total stands at more than 150. The technology is constantly evolving. The current generation of instruments is smaller than earlier models, and the tags are now deployed with an air-powered applicator rather than a crossbow; they're implanted with only the antenna and saltwater switch protruding. To guard against infection, the tags also contain a time-dispersed antibiotic that lasts up to eight months. All satellite tags eventually stop transmitting: either they malfunction, the batteries die, or the whales work them loose like slivers. These tweaks in design, however, have meant that the tags are staying on for much longer than in the past—the current record for blue whales is 495 days. Someday, with enough long-term data, and with repeat deployments on the same animal in different years, the picture of the blue whale's peregrinations will become a whole lot clearer.

As Calambokidis, Sears and other researchers started to answer some of the population-scale questions about blue whales—how many there are, where and when they migrate—they also began to think about the smaller picture, too. "We started to get interested in understanding the behaviour of individual blue whales," says Calambokidis, "and we started to look for ways to get an insight. We see these animals for just a few seconds at the surface, and then they disappear for 10 or 15 minutes at a time and we have no idea what they're doing underwater. How can we learn about what their world is like and what's going on down there?" Here again, it was tagging that provided a window.

Most people do not appreciate just how little scientists understood about whale behaviour until recently, though one can hardly blame the biologists for the lack of knowledge. Scientists are adept at studying land animals they can observe up close for long periods, even if the animal's habitat is remote and miserably uncomfortable. They have hunkered down in Antarctica to study the mating habits of emperor penguins and waded through Amazon tributaries to learn about parental care in poison dart frogs. Yet we simply do not have the means of observing animals underwater for any significant length of time. Whales offer glimpses when they surface to breathe, but scientists are left to guess at the rest. Sears, who grew up reading the work of the pioneering biologists who studied African mammals in the 1960s, offers a comparison: "Imagine doing that work in the Serengeti if it was blanketed with a thick fog, so all you could see was the nostrils of giraffes. What those people could do in one or two years takes us 20 or 30."

It's not just a question of depth. While blues can be seen within sight of the coast in some areas, enormous portions of their lives are spent far offshore. It is impossible to get a complete picture of a population if you are limited by the range of an outboard motor. Of course, whale biologists can and do get hundreds of miles offshore in large vessels, but these expeditions are absurdly expensive. The ships needed to get to areas such as the Costa Rica Dome in the Pacific, or the Flemish Cap in the Atlantic, can cost of thousands of dollars a day to lease and maintain, and no government is willing to pony up that kind of money. Indeed, whale researchers need to have a certain entrepreneurial streak, and it doesn't hurt to have benefactors. Richard Sears, for example, has led research cruises in the Atlantic using a vessel owned by a whale lover who made her fortune as one of the creators of the board game *Trivial Pursuit*. And much of Bruce Mate's offshore work is done from a refitted 83-foot trawler donated by a west coast fishing family. "There just isn't the will and the mandate within government agencies to see some of these jobs done," Mate says. "So it's going to have to be left largely to the private sector."

Another obstacle is the physical and psychological demands of long-term whale research, which involves months on lonely roads and angry seas. "You sacrifice relationships and a lot of things," says Sears, who has never married or had children. "I've found that women don't like to hang out with people who don't make a lot of money and who disappear once in a while." He's also sacrificed his back and knees to the sea gods, as have some of his colleagues. It takes a special kind of dedication to do this work. Bruce Mate, after aggravating an injury caused by bouncing around in boats, once remarked, "Sometimes I remind myself that if this stuff was easy, somebody would already have done it and know the answers."[1]

The fog started to lift in the late 1990s with the appearance of new tags called recoverable data loggers. Unlike the satellite tags that had already been used on blue whales, these are not designed to stay put for months and track the location of whales over great distances. Instead, data loggers are attached with suction cups that rarely hold for longer than a few hours. In that short amount of time, they can measure and record the depth to which a whale dives, the salinity and temperature of the water, and the speed and direction the animal is swimming. Many can also record the whale's vocalizations, and some include underwater cameras. All of this makes the instruments fundamentally different from satellite tags, which do not need to be recovered. The tags used to study behaviour collect far more information than can be transmitted to a remote computer, and instead store their data on a memory chip, which means the scientists must find the tag after it falls off. The two types of instrument complement one another with a mix of low and high resolution: satellite tags provide a little information (location only) for a long time, while behavioural tags provide a lot of information compiled in a much briefer period.

As with so many technological advances, this one was inspired by nature. In 1986, the same year Calambokidis first encountered blue whales in the Gulf of the Farallones, a graduate student in marine biology named Greg Marshall was on a scuba-diving trip in Belize. On one of his dives, he noticed a remora, or suckerfish, hitching a ride on a passing shark. He had a sudden brainwave: what if we could attach a video camera to a shark? What would we learn if we could see what that remora sees? Think about the bulky camcorders people carried around in the 1980s and you will appreciate the challenges Marshall faced. Was it possible to make a camera so small that it would not interfere with the animal's behaviour? And if it was, how would you attach it? To be of any practical use, the camera would need to stay on for several hours, then somehow detach itself so it could be retrieved. In 1991, the National Geographic Society saw promise in the technology and funded its development. Eventually it became known as the Crittercam, and it provided some of the unique footage seen by millions in the popular film *March of the Penguins*.

In 1999, Calambokidis joined National Geographic scientist John Francis and others to see what the Crittercam could reveal about the underwater behaviour of blue whales. The instrument package included a modified Hi-8 camera, sensors to measure depth and temperature, and a hydrophone to record the whales' vocalizations, all inside a casing the size of a shoebox. The first challenge was simply figuring out a way of attaching something that large to a blue whale. Their solution was to equip the Crittercam with a silicone suction cup, about the size of a salad plate, and mount the whole thing on a long, counterweighted pole. One team member would drive an inflatable alongside the whale while Francis reached out with the pole and placed the Crittercam on the animal's back as it surfaced. A vacuum pump instantly removed the air from the suction cup, sticking the camera to the whale's skin. As for the problem of getting it back, they fitted the suction mechanism with a piece of magnesium designed to corrode after about three hours in seawater. That broke the seal and caused the Crittercam to detach and float to the surface. A VHF transmitter then sent out a signal that allowed the researchers to find and retrieve the floating instrument.

This method obviously requires some expert skill at handling a boat. Blue whales are fast swimmers and surface for only a few seconds at a time, and the Crittercam couldn't be slapped on just anywhere; ideally, it would be positioned high on the back, well forward of the dorsal fin. During their first efforts in Bodega Bay, California, in the late summer of 1999, Calambokidis and his team approached at least 10 whales, making contact with seven of them, but they attached only one Crittercam. They tried again the following year in Monterey Bay and successfully tagged the lead animal in a pair. The magnesium turned out to be hardier than expected and the device stayed put for more than 10 hours, with the scientists following along until bad weather and darkness forced them to return to shore without the Crittercam. It took three days of patiently following the VHF signal before the researchers finally found it bobbing in the sea, more than 25 miles from where they had attached the camera. Moisture had condensed inside the casing, causing some loss of data, but it was a promising start.

In February and March 2001, Calambokidis was in the Sea of Cortez for another round. The first camera the researchers attached stayed put for more than six hours and was successfully retrieved, and two days later they attached a second one for several hours, but were unable to find it after it fell off. With a few adjustments to the equipment, the team was now finally getting the hang of it, so when they visited the Santa Barbara Channel that July they looked forward to more success. When they began their search, however, they initially found almost no blue whales. Then came one of those days when biologists must feel that heaven is smiling on them. Scouting off San Miguel Island, about as far from a harbour as you can safely go in an inflatable, they happened upon one of the biggest concentrations of blue whales Calambokidis has ever seen: more than 200 animals spread over about 20 square miles. Amid such a bounty, they made 12 successful deployments of the Crittercam.

As remarkable as technology is, watching video shot from the backs of blue whales is a bit disappointing. The picture is dark (reddish LEDs occasionally illuminate things close by) and little of the whale's body is ever visible. Certainly it's no *March of the Penguins*. But the Crittercam's sensors did make some important discoveries about the whales' undersea behaviour. For example, the whale they tagged in the Sea of Cortez in March 2001 turned out to be feeding, which in itself was interesting, since the primary feeding area for blue whales in the eastern North Pacific was believed to be off California, where they tend to travel later in the year. As Calambokidis and his team followed the tagged whale in their boat, the echo sounder picked up a large patch of krill between 200 and 300 feet below the surface. When they later retrieved the Crittercam and uploaded the data from the depth sensor, they noticed something dramatic: during the first couple of hours the instrument was attached, the whale made a series of deep dives below this layer of krill, and each dive was followed by several upward lunges. That was a radically different feeding strategy from the earlier model that had blue whales swimming horizontally through the krill at a more or less steady speed. "In this case," says Calambokidis, "we saw very little, but we learned a crucial thing: the whale came head-up and darted into the krill layer from underneath. We got a good picture of the way whales were approaching the krill. Before this, people didn't really understand how that was occurring."

This same deployment also gave the scientists a look at how a blue whale's diving pattern changes as darkness falls. Krill tend to move away from light, so as evening falls they migrate toward the surface, and the depth sensor on the Crittercam showed that the whale followed the same pattern. After about 6:30 p.m., the animal's dives became progressively shallower and the pattern of upward lunges ceased. As 9 p.m. approached, the whale was staying down for only a minute at a time, rarely going below 65 feet. This series of languid surfacings may be the blue whale's method of sleeping. Studies of other cetaceans have shown they do not breathe unconsciously as humans do, but instead rest one hemisphere of their brain at a time, while the other remains alert enough to control movement and respiration.

Another surprise was how rare it was to see other whales in the video. Calambokidis thought that if he could place a Crittercam on one of the animals in a pair, the other might be visible underwater;

perhaps it would provide the first footage of a blue whale feeding below the surface. In fact, the pictures almost never showed a second whale. "We began to realize that, unlike with humpbacks, it doesn't appear that blue whales are cooperatively feeding. Even though they often pair up, they still appear to feed apart from each other." No one had observed blues engaged in this cooperative behaviour near the surface, but scientists wondered whether they might be doing something similar at depth. These early data did not support that idea.

A few fascinating seconds of video did come from one of the deployments Calambokidis made among the blues off San Miguel Island. The scientists knew that many blue whales in one area must indicate vast clouds of krill. But their boat's echo sounder wasn't showing much in the first 650 feet of water, and blues were believed to do most of their feeding within that depth. So where was all the food? Their question was answered when they watched the Crittercam video: as the tagged whale dove, tiny krill streamed past the camera like wind-driven snow, getting more and more dense until the hungry animal finally began to lunge. It appeared that the whale was ignoring the diffuse krill layer near the surface, preferring to wait until it reached the motherlode below. "These whales were diving down to over 300 metres," Calambokidis explains, "deeper than we thought blue whales go to feed."

While the Crittercam video has certainly provided useful data, researchers now rely less on pictures and more on data collected with other instruments. Indeed, most of the data loggers that researchers now place on the backs of blue whales have no video component. There are many models, and researchers can custom-build them to suit their needs, but making one that works reliably after being dunked in seawater a few hundred times is more difficult than it may seem. One of the most successful instruments used on blue whales is a small torpedo-shaped device known as a bioacoustic probe, or B-probe for short, built by a California company called Greeneridge Sciences. At the heart of this marvellous little piece of engineering is a hydrophone and a solid-state memory chip that records vocalizations. The tags, which cost several thousand dollars each, also house a package of other instruments, including a time-depth recorder, a sensor to measure water temperature and an accelerometer that records the whale's pitch and roll angle. Best of all, unlike the bulky Crittercam, the B-probe is just 8 inches long, weighs 7 ounces, and all of its instruments are potted in a sturdy, waterproof resin. Its creator boasts that you can drive a nail with it.

Scientists began attaching early versions of bioacoustic probes to marine mammals in the mid-1990s, making their first efforts with northern elephant seals in California. Like the Crittercam, however, the B-probe presented challenges that needed to be overcome before it would work with blue whales. It is easy enough to make the probe float by slipping it inside a foam collar, and attaching a VHF transmitter allows the scientists to track the device while it's on the whale and after it falls off. One of the more difficult problems, however, was coming up with a way to attach it to the whale. Because the B-probe is so much lighter than a Crittercam, scientists began experimenting with much smaller suction cups that could simply be pressed onto the back of the animal, without requiring an awkward vacuum pump. They soon found that two cups, one behind the other along the length of the tag, held the tag snugly, but there was still the problem of releasing it after the data were collected. The dissolving magnesium was impractical, so the researchers experimented with some decidedly low-tech methods. They tried making a hole in the suction cups and inserting various substances—including Tylenol tablets and bits of throat lozenge—figuring they would eventually dissolve and break the seal. One of the more promising suggestions was to use Gummi Bears, since the candies are small and pliable. ("The main problem was that John kept eating them," jokes one of Calambokidis's colleagues.) But none of these substances readily dissolved, and sometimes they flew out of the hole as soon as the tag made contact with the whale. In the end, the scientists resigned themselves to following behind until the tag decides to come off on its own. Generally, it stays put for a few minutes to two days, with the scientists considering anything longer than a couple of hours to be a success.

The data from one early deployment of the B-probe helped Calambokidis and his team paint a clearer picture of what blue whales do when they dive into a krill layer several hundred feet below the surface. For example, at the start of one 13-minute dive sequence, the tag showed that the whale descended to almost 1,000 feet at an angle of about minus-80 degrees; that is, pointed almost straight down. "These little wiggles here are actually fluke beats," Calambokidis explains as he points to a graph of the data. "Every time the whale beats its flukes, it creates this oscillation in the pitch sensor. So you'll notice that it was beating its flukes when it was at the surface, but then it put its head down, did one beat, and then largely just glided down. Once it reaches a depth of around 65 or 100 feet, it becomes more and more negatively buoyant, and from our measurements, we can see it's actually accelerating on this descent, even though it's not beating its flukes. It gives a whole other perspective—for an air-breathing mammal to sink like a rock is kind of scary." It also demonstrates that while negative buoyancy can help blue whales conserve oxygen as they dive, they pay for it on the way back up.

After its descent, the whale then made a series of upward lunges into the krill, creating a sawtooth pattern on the graph, and the tag's pitch sensor revealed that the animal was rolling to the side as it came up into the prey layer. Observers had long noticed that blue whales tend to turn themselves sideways, even completely upside down, as they feed near the surface. The tag data showed for the first time that they were performing similar acrobatics at depth. Every 20 or 30 seconds, the tagged whale would angle itself upward, beat its flukes three or four times and then lunge, open-mouthed and silently, into the krill. Then it pointed itself almost straight up and beat its flukes madly to get back to the surface.

Once whale researchers were able to deploy tags reliably, it created new opportunities to learn about other poorly understood aspects of the animals' underwater lives. One of these mysteries involved why blue whales spend relatively little time submerged when they are feeding. In general, the larger a marine mammal's body, the more oxygen it can store, and therefore the longer it can remain underwater. Based on an animal's mass and metabolic rate (the rate at which it uses up its oxygen stores), biologists can calculate its "theoretical aerobic dive limit"—that is, the maximum length of time that the animal should be able to stay submerged before being forced back to the surface to breathe. For blue whales, this theoretical limit is over 31 minutes, but in practice, their foraging dives average less than 8 minutes. Why are they staying down for only one-quarter of the time their oxygen stores should permit?

In 2001, three California-based scientists—Alejandro Acevedo-Gutiérrez, Don Croll and Bernie Tershy—set out to tackle that question. Their idea was to measure how many lunges blue whales made during a series of dives, and then to compare this with how long it took the animals to recover at the surface. They expected that the greater the number of lunges, the longer the whale would subsequently spend breathing, just as a man would need more time to catch his breath after running up four flights of stairs compared with two. If this prediction turned out to be true, it would help explain why a blue whale's feeding dives are relatively short: the tremendous energy cost of lunging forces them to come up for air more quickly. To test this, the scientists would need hard data, which they got by attaching time-depth recorders to the whales. Rather than being held on by suction cups, these tags, about the size of a large flashlight, were shot from a crossbow and secured to the whale's skin by an inch-long barb. To solve the problem of releasing the tag, engineers built a tiny receiver that could pick up a signal from the researchers' boat. When they felt they had enough data, the scientists pressed a button, the signal activated a cutter that snipped the wire connecting the tag and the barb, and the instrument floated to the surface.

The team successfully deployed and recovered the tags from seven blue whales off Mexico and California. The animals descended to an average depth of 430 feet and made up to four lunges during each dive. The speed of the whale was much greater during the upward thrust of the lunge than during the downward movement, which indicates that the animal was beating its flukes hard in order to propel its massive body against the pull of gravity. And gravity isn't the only force the whale must contend with. When a blue whale lunges into a krill patch, its mouth opens almost 90 degrees and may engulf 70 percent of its body mass in seawater, or somewhere in the vicinity of 65 tons. The hydrodynamic drag created by this process is staggering; it is enough to overcome tremendous inertia and bring the lunging whale to almost a complete stop. Clearly, all of this involves strenuous effort on the whale's part. In fact, the scientists calculated that the energy cost of these feeding lunges was about three times what the whale expends during a non-feeding dive. This helps explain why the whales are staying down for much less time than the 31 minutes predicted by the model. To return to our human analogy, a man standing motionless might be able to hold his breath for a minute, but he would likely be unable to do so while running up a staircase.

Satellite tags and data loggers have done for blue whale research what the microscope did for biologists, or the telescope for astronomers, and every year, the technology evolves, improves and promises more. In 2007, Bruce Mate began experimenting with a new tag that uses GPS technology designed to get a fix on the satellites during the few brief seconds that the whale was at the surface. There are still kinks to work out, but these may eventually allow him to plot the path of travelling whales with remarkable precision. Future generations of the B-probe will answer more questions, too: more sensitive temperature sensors may provide clues about how blue whales find krill, and a three-dimensional accelerometer that can record the pitch, roll and yaw of the whale may reveal more about its feeding strategies. For blue whales and the scientists who study them, the game of tag is just beginning.

EIGHT

BLUE NOTES

CRITTERCAMS, time-depth recorders and other tags have revealed much by acting as the underwater eyes of scientists. But many blue whale behaviours simply cannot be decoded by watching the animals, either directly or remotely. To understand blue whales, it's just as important—perhaps more so—to listen to them. Sound is critical to the lives of blue whales, and many scientists now believe that acoustic research—a field that exploded in the early 1990s—may turn out to be the single most useful tool for understanding the world's grandest animal.

Natural selection favours adaptations that give individuals some survival benefit, and keen vision does not offer any great advantage to a marine mammal. Light levels drop about 90 percent for every 250 feet of depth, even in the clearest conditions, and in plankton-rich waters visibility can be poor even at a few feet. In most situations, even the most eagle-eyed blue whale would probably be unable to see much beyond a single body length, and an animal feeding on krill at 500 to 600 feet is essentially lunging in the dark. What about taste and smell? Baleen whales do have more chemical receptors around their blowholes than do toothed whales, and it is possible that they are able to sniff the air when they breathe at the surface; some research has even hinted they may catch whiffs of the dimethyl sulphide emitted by phytoplankton. But it seems unlikely that smell is of much use to blue whales or any other cetacean. Their sense of taste is not well understood either. Some biologists believe that whales are able to pick up cues from the urine or feces of other individuals, which may linger in the water after the animal that produced it has swam off. Blue whales may also use taste to assess the concentration of krill patches, holding off opening their mouths until the prey is densest, though this has never been demonstrated. In any case, vision, smell and taste are useful to them only over relatively short distances. All the evidence points to hearing as their most important sense.

Sound travels about five times faster through water than through air—about 5,000 feet per second versus 1,000 feet per second, though the exact figure varies. In air, the only significant factor is temperature: sound travels more quickly through warm air. In the ocean, however, the speed of sound is affected by temperature, pressure and, to a lesser extent, salinity; increasing any of these causes sound waves to move faster. The differences are significant enough that oceanographers can measure these properties over large areas by sending out sound pulses and monitoring their changes in speed.

The ocean, of course, is not homogeneous. Temperature and pressure vary greatly with depth, creating a series of horizontal layers. (In an estuary such as the St. Lawrence, salinity also varies considerably.) At the surface, there is generally a warm layer, where pressure and temperature don't change dramatically with depth. Beneath this is an intermediate level called the thermocline, where temperature decreases sharply as you go deeper, while pressure increases only moderately. Finally, in the so-called deep isothermic layer, the temperature is consistently cold, but the pressure continues to increase with every foot. Sound waves travelling through the sea always bend toward the layer where the speed of sound is at a minimum: those in the thermocline bend toward the seafloor, where it is colder, but when they reach the constant temperature of the very deep ocean, they bend upward because of the increasing pressure. Then, on reaching the thermocline again, they bend back downward and the process starts over. The net effect is that the sound moves more or less horizontally along a pipeline between the bottom of the thermocline and the top of the deep isothermic layer. This pipeline is called the deep sound channel (or sometimes the SOFAR channel, the acronym derived from "sound fixing and ranging"), and at low and mid-latitudes it usually occurs some 1,900 to 4,000 feet below the surface. Refraction causes no loss of energy, so sounds bending at the top and bottom of the channel can travel for many miles. In extreme cases, they can even be heard across entire ocean basins. Air guns fired under the sea during surveys off California, for example, have been detected by instruments in Polynesia, more than 3,700 miles away.

The other critical point to understand about ocean acoustics, at least as it relates to blue whales, is that low-frequency sounds travel farther than high-pitched ones. All sounds in the sea lose energy when they are scattered, reflected and absorbed by particles in the water, as well as by contact with the surface and the seafloor. But low frequencies are much less susceptible to these losses because of their longer wavelengths. A 20-Hz tone—which, as we will soon see, is a very low frequency within the range of blue whale calls—has a wavelength of over 250 feet in water. Sound waves of this size are not much affected by small particles, and they lose far less energy when they bounce off surfaces, which means they must travel great distances before the losses add up significantly.

As for the ability of blue whales to detect sounds—especially at very low frequencies and across great distances—not much is known for certain. However, animals tend to evolve ears that are most sensitive to the frequencies of sound emitted by others of the same species. (Or perhaps the other way around: they may have evolved vocalizations to match the frequencies they hear best.) Human hearing, for example, is most sensitive to sounds of about 1,000 to 3,500 Hz, which is typical of the frequencies in our speech. Blue whale ears, then, can almost certainly hear very low sounds, even below 10 Hz, since these are the frequencies they use to vocalize. The animals probably determine the direction a sound is coming from the same way we do: by detecting the tiny difference between the time it arrives in one ear and the time it reaches the other. This is called the sound's phase shift, and the principle is similar to the way we use both eyes for depth perception. Determining the angle of the sound's origin is easier when one's ears are farther apart, since the greater separation increases the phase shift, and in blue whales the left and right ears can easily be 15 feet apart.

As for the mechanism blue whales use to emit sounds, these are a complete mystery. Calling whales have been observed by scientists, but the animals give no visual clues. They do not open their mouths, and while they can make noisy exhalations through their blowholes, these sounds are unrelated to their vocalizations, which originate somewhere within the body. Presumably they are generated by moving air through some unknown valve, with another unknown structure acting as a resonator. Whatever is going on inside this biological bagpipe, the long, low sounds of blue whales would displace an astounding 800 to 1,100 litres of air, according to one estimate. That's about half the amount the animal is able to store in its lungs near the surface, though these organs would be under tremendous pressure during a dive. Some researchers have suggested that a calling whale might actually need to dive or ascend to allow the pressure of the ocean to compress and relax its lungs like the bellows of an accordion. However the calls are made, scientists agree that they spread out from the animal's body in a rough sphere. This is different from other species. The heads of toothed whales, for example, have fat-filled cavities called melons, which they can reshape in order to direct and focus their echolocation clicks. Even if baleen whales had such melons, they would be ineffective on low-frequency sounds. Remember, a 20-Hz sound wave in water is about 250 feet long. It's hard to see how any structure, even in a blue whale's body, could be large enough to act as an acoustic lens for that enormous wavelength.

If scientific knowledge about the blue whale's acoustic world is sketchy now, it was nil until the late 1960s, and even then the only people who knew anything about the sounds made by large whales were in the U.S. Navy. After the Second World War, the navy constructed a vast network of hydrophone arrays, some towed by vessels and others positioned on the seafloor—the locations are still secret—in the Atlantic and Pacific. The role of this Integrated Undersea Surveillance System (IUSS) was to listen for Soviet submarines, but the arrays also picked up signals that looked nothing like enemy subs. The sounds that raised the most eyebrows were the pulses around 20 Hz, which is at the extreme low end of the human hearing range. These signals were very regular in their patterns, which suggested they were not random seismic rumblings. They were also extremely intense—so powerful that some listeners attributed them to military activity or blamed the signals on faulty electronics, though other navy scientists suggested they were "biologicals." At least a few speculated the sounds were made by large whales, possibly for the purposes of navigation; that is, the whales might be bouncing pulses off continental shelves or seamounts to help them fix location and direction.

The hydrophone arrays detected countless sounds of this nature before the 1970s, but the data were classified and unavailable to civilian scientists. In any case, it wasn't yet clear exactly what the sounds were. There wasn't anything to compare them with, since precious few recordings were made by shipboard scientists who could actually see and identify the animals making the calls. Even when researchers attempted this, they quickly discovered one of the biggest challenges in whale acoustics, which continues today: when you make a recording in the presence of several whales, it's extremely difficult, sometimes impossible, to know which one is actually emitting the sounds. The vocalizations are often made by animals that are never seen at all. During one early attempt in May 1969, two Canadian researchers in the North Atlantic encountered a feeding blue whale and lowered a hydrophone into the water to record its vocalizations. They described these as "clicks," including some that were too high for human ears to detect, and speculated that blue whales might emit ultrasonic pulses to echolocate krill the way bats do with insects. It later became clear that these sounds were not coming from the blue whale, but from unseen porpoises nearby.

The first shipboard recording to capture what really were blue whale calls was made a year later by William Cummings and Paul Thompson, two navy scientists on a research cruise off Chile. The sounds were three-part patterns of what Cummings and Thompson called "low-frequency moans," which lasted more than half a minute. Almost all of these moans were below 200 Hz (lower than the G below middle C on a piano), and the strongest were between 20 and 32 Hz. More remarkable, however, was the sheer power of these vocalizations. "The energy of the blue whale's signal is spectacular," Cummings and Thompson wrote, noting that these were at the time the most powerful sounds known from any living creature, averaging 188 decibels, "the same as the overall source level from a World War II USN cruiser at normal speed." The scientists, clearly impressed, were moved to add an elegiac note in their paper: "This finding is especially noteworthy because it is doubtful if these animals, the largest ever to inhabit the earth, will survive man's overharvest."

Before discussing blue whales and their mighty vocalizations further, a few words are in order about how sound levels are measured. The decibel (dB) is not a linear unit like the mile or pound. Rather, it is a way of describing the ratio between the measured pressure of a given sound and a fixed reference. Scientists use the quietest sound the typical human can hear—a sound that exerts a pressure of 20 micropascals—as the fixed reference in air. In other words, a sound that is right at this threshold is said to be 0 dB, which simply means that the intensity of the sound is zero times higher than the reference. The scale then increases according to powers of 10, so 10 dB is 10 times more powerful than the reference, 20 dB is 100 (10^2) times more, and 30 dB is 1,000 (10^3) times more, and so on. Many people mistakenly believe that 96 dB isn't much more than 90 dB—in fact, it's twice as powerful. To further complicate things, scientists use the 20 micropascal reference in air because it is relevant to human hearing, but it makes little sense to use the same benchmark for sounds in the ocean. When measuring underwater sound levels, therefore, they use a reference pressure of 1 micropascal at a distance of 1 metre. And because water is denser than air, the sound energy needed to exert a micropascal of pressure is different in the two mediums. That means that 96 dB in air and 96 dB in water are apples and oranges.

These misunderstandings have caused many people—including some who should know better—to make claims about blue whale sounds that are at best misleading, and at worst wholly incorrect. Undersea earthquakes can measure about 220 dB, leading one well-known naturalist to remark that this "isn't all that different" from a 185 dB blue whale call.[1] In fact, there's an *enormous* difference between those two levels: blue whale calls are nowhere near as powerful as large earthquakes. It is equally fatuous to say, as books and articles often do, that blue whales are as loud as a rock concert or a jet engine. There is simply no way to describe how loud a calling blue whale would be above the water, not least because many of the animals' most energetic sounds are at frequencies that humans cannot hear. The question starts to sound like the Zen koan about one hand clapping: how loud is an inaudible underwater sound when it is travelling through air?

After Cummings and Thompson made their breakthrough recording, they realized that many of the long, low-frequency pulses the navy had been hearing in the Pacific for years were indeed the calls of blue whales. Two others working in the North Atlantic, William Watkins and William Schevill, had already made a similar discovery, attributing the shorter 20 Hz signals to fin whales. These navy scientists published some of their findings, but most of their data remained classified. Meanwhile, others were left to guess what these low-frequency blue and fin whale sounds might be used for. The most common ideas were that they were used for navigation, echolocation, or for communication with nearby animals, although at least one researcher proposed that the pulses might be the whales' heartbeats.

Then came a revolutionary idea from a maverick scientist—one of those insights, like plate tectonics and black holes, that is initially met with derision but turns out to be way ahead of its time. It arrived in the form of a 1971 paper in the *Annals of the New York Academy of Sciences*, one of the oldest scientific journals in the United States. The article's lead author, Roger Payne, had already made a name as a pioneer in whale research. During his work among right whales in Patagonia, he invented the photo-identification technique that would later become widely used with other species, and his early recordings of humpback songs helped transform the way the public thought about whales. Now he, along with ocean-acoustics expert Douglas Webb, turned their attention to those low-frequency pulses that military hydrophones continued to detect throughout the Atlantic and Pacific. Payne and Webb were not the first to assert that these were probably made by baleen whales, but they took the notion a giant leap further. They argued that fin whales, and probably blues as well, "may be in tenuous acoustic contact throughout a relatively enormous volume of ocean and that such contact might be of use for finding each other or for joining, or keeping together in, widely dispersed herds."[2] Payne was quick to clarify that he wasn't suggesting that the animals were having conversations or sharing complex information, but might simply be sending out a kind of locator beacon.

He went on to describe how the mysterious signals being picked up by navy hydrophones had characteristics that made them ideally suited to whales communicating across vast stretches of ocean. First, there was their intensity; if these were indeed whale calls, they were more powerful than any sounds known to be emitted by an animal, and therefore quite capable of traversing great distances, especially if they were getting into the deep sound channel. The second factor was their low frequency. As we've seen, sounds centred around 20 Hz would travel much farther than the higher-frequency sounds associated with dolphins and toothed whales. The 20-Hz figure was actually something of a magic number, Payne and Webb suggested, falling right between the frequency of the earth's deep grumbles and the sound of turbulence caused by storms, both of which might otherwise mask the signal of a calling whale. Natural selection might well have favoured 20 Hz because it is "the highest frequency one could employ to build a long-range signalling system free from all weather noise"[3] except the strongest storms, and the lowest frequency whales could use without interference from seismic activity.

The sounds heard by the navy scientists were also repetitive: one signal heard repeatedly in the North Atlantic was a 1-second pulse, followed by a 12-second pause, then another pulse, with the pattern repeating for about 15 minutes, after which there was silence for about two and a half minutes. This timing, Payne noticed, fits with the cycle of a whale's breathing and diving. More tellingly, a repetitive signal like this would not require an active listener. As long as another whale eventually swam within range, the repeated signal would get noticed, even if it took hours or days. Think of it this way: if you fall down a well, your best strategy is not to shout for help once, but to repeat the same distress call over and over to maximize the chance that a passerby will notice it and come to your rescue.

The purity and duration of the sounds were meaningful for Payne as well. The signals were centred around 20 Hz, and the range of frequencies—what acousticians call the bandwidth—was only about 3 Hz. A predictable, redundant signal like this is easier to detect amid the background noise of the sea. A listening whale would be able to concentrate on that narrow bandwidth around 20 Hz. Human experiments also show that a 1-second signal is much easier to detect than a very brief tone, but a 3- or 4-second sound is not appreciably more noticeable. The duration of the pulses attributed to fin whales were almost exactly 1 second, which means they would provide maximum benefit to the whale with no wasted energy.

Payne also pointed out that animals using this form of communication would have to be extremely large. Only an immense animal would be able to emit calls at 185 dB, and one would need widely spaced ears to fix the direction of a distant call. Finally, only a creature so large that it had little or no fear of predators would be able to safely use a long-range communication technique that would alert others to its presence.

Tying together all of these threads, Payne made a startling hypothesis. He suggested that fin and blue whales organize themselves in social units he called "range herds," in which "the population lives in tenuous contact throughout large portions of its range."[4] The powerful, low, monotonous and repetitive sounds that navy scientists had been hearing for years, he proposed, may be whales that are hundreds, even thousands of miles apart, periodically signalling their location to others. Noting that scientists who tried to track the seasonal movements of blue whales were (and still are) repeatedly confounded, Payne wondered how the whales find each other to mate. He offered an explanation: "I believe the reason that whalers have never found a breeding ground for blue and fin whales is because none exists. I suspect that when blue and fin whales wish to find a mate—when 'he' wishes to find a 'she'—they simply start calling. Even if it takes several weeks of swimming to get there, getting together could be achieved by homing on the sounds of individuals or groups of whales."[5] If Payne is right, it would mean that even though blue whales do not travel in large pods the way dolphins and many smaller whales do, their social groups may radiate outward over enormous areas, with individuals staying in touch through their own brand of long-distance calling.

Payne and Webb's estimates of how far whales might be signalling irked many of their contemporaries. ("I suspect my career was held back significantly by the way people responded to the theory," Payne would later write.[6]) Assuming the oceans were free of background noise and that the calling whale was right inside the deep sound channel, they estimated that 20-Hz sounds might carry up to 11,500 nautical miles. They readily admitted that this was only a theoretical upper limit that would never be reached, and under pressure from navy scientists such as William Cummings, who considered this idea ridiculous, Payne and Webb revised their estimate downward, suggesting that the sounds might propagate about 500 miles. As Payne delivered his stirring talks all over the U.S., however, many laypeople were left with the impression that the whales were talking to one another across the globe. Unfortunately for everyone interested in exploring the question further, precious few blue whale recordings were made over the next two decades, except for those cloaked in military secrecy.

Then, in August 1991, a coup attempt in the Soviet Union set in motion a series of events that would end months later in the collapse of the communist superpower. This event had a profound effect on blues as well as Reds. With the Cold War over, the U.S. Navy's expensive submarine-sensing hydrophones became less important, and their data less politically sensitive. Within a year, Democratic senators such as Al Gore and Ted Kennedy began to suggest that the military consider allowing civilians to use the equipment for science. The navy agreed and invited biologist Chris Clark to be the lead researcher on a project they called Whales '93. Clark was head of the Bioacoustics Research Program at Cornell University in Ithaca, New York, and along with a team of the navy's own specialists, his task was to see how the network of instruments might be used to study whales. What species would the hydrophones be able to detect? Could individual whales be tracked like submarines? Could the arrays be used to monitor the seasonal movements of populations?

It's hard to overstate what a godsend this was for non-military scientists interested in baleen whale acoustics. A dearth of data became an embarrassment of riches almost overnight. On the first weekend Clark had access to the arrays, in November 1992, they detected so many calls from blue, fin and minke whales that their database crashed and they had to design a new one that could handle the avalanche of signals. "During the very first hours that the IUSS was used to detect blue whales," Clark recalled, "more of their sounds were recorded than had ever been described in the entire scientific literature."[7] (These sounds had, of course, been described for decades in classified military documents.) In the ensuing months he learned that many of the blues he was hearing were in the region of the Grand Banks, which Richard Sears had long suspected was a winter destination for at least some of the St. Lawrence blues. The calls were most frequent between September and February and tapered off by late May and early June, which is just what you'd expect if these were the same animals.

Clark recalled the elation he felt the first time he detected a blue whale calling. Inside a concrete building in Virginia, where the hydrophone data were streamed in real time, he was listening to an animal somewhere off Newfoundland, though the calls were so low he could barely hear them in his headphones. He watched with excitement, however, as yards of paper rolled off the analysis machine with a constant string of comma-shaped spectrograms that were undeniably the calls of a blue whale. He glanced at the name of the hydrophone detecting these sounds and then looked for its location on his wall map of the Atlantic. "I froze," he recalled, "because I realized that the whale whose voice was responsible for these printed patterns and the faint regular hum in my ears was about *500 miles* away!"[8] As he continued his work, detections from that distance turned out to be routine. Whales calling off Newfoundland, in fact, could be detected all over the western North Atlantic, down to the West Indies. As the realization sunk in, Clark thought of the one man who would be most vindicated by the discovery that whales could be heard across vast stretches of ocean. Clark had been a graduate student under the supervision of this scientist in the 1970s and had helped with his research on right whales off Argentina, but the two had since grown apart. In June 1994, however, they met up again in an old churchyard in England. As the men sat down on the bank of a stream and Clark shared his discovery, the rapt listener was none other than Roger Payne, who wrote about the moment shortly afterwards:

> When Chris Clark said to me, 'Roger, I have tracked a blue whale from over a thousand miles away,' the feeling that came over me is one I may never recapture . . . For me, however, the most moving thing about the moment was that it was Chris, the prodigal son, who had gone out and done what I could not do

and was coming back in triumph to tell me. A feeling of amity swept over me, bringing me almost to tears.9

The IUSS was designed not merely to detect the presence of enemy submarines, but also to follow their movements. In theory, the same process, known as adaptive beamforming, can be used to track individual whales, but it soon became apparent that almost all the blues in a given area sounded similar, so zeroing in on individuals was extremely difficult. If a whale had a particularly unusual call pattern, however, it was possible. Clark and his navy colleagues tracked one animal, nicknamed Ol' Blue, for 43 days as it roamed over 2,700 nautical miles in the vicinity of Bermuda. Another whale in the North Pacific has been tracked every year since 1992. Its unique call, centred at 52 Hz, is unlike any known baleen whale vocalization. Perhaps it is a hybrid, or simply an individual blue or fin whale with some physiology that gives it a higher voice. Either way, it conjures that image of a lonely animal calling and calling in a vast ocean without ever being acknowledged by another.

Access to the navy hydrophones in the early 1990s did more than just provide a motherlode of data. It also fired the imaginations of a generation of scientists who saw an opportunity to revolutionize the study of large whales. One of the young scientists who found herself part of this movement was Catherine Berchok, who first went to work with Richard Sears in the St. Lawrence in 1992. While she learned a great deal with the MICS team, she quickly realized that observing the surface behaviour of whales wasn't enough to understand them. "You sit there in the boat, and if you're lucky you get a whole minute with the whale on the surface before it dives, and then it's a half hour before it surfaces again. It just got really frustrating. They live most of their lives under the water, and they're not using vision most of the time. That's why I got interested in acoustics." While pursuing her PhD at Pennsylvania State University, Berchok returned many times to the St. Lawrence—now with hydrophone in hand—to collect data for her thesis, and perhaps to help answer some of the questions Sears had been asking for over a decade. "My goal in the beginning was to be able to plug into the navy data and say, 'OK, here's a certain whale. Can you tell us where it goes in winter when it leaves the St. Lawrence?' During the time I was doing my acoustic work, Richard was trying this with satellite tags, but most of them failed. And even if the satellite tags work, you can only tag one whale at a time, whereas as long as you have a hydrophone in the water, you get information on whoever is vocalizing as they go by."

Acoustics promised to overcome many of the perennial challenges faced by field biologists like Sears and Calambokidis. When it was too rough to go to sea, researchers imagined themselves sitting at their computers with a cup of coffee and listening for the whales in their study area. The mysterious offshore movements of blues might be tracked with hydrophones deployed all over the world's oceans. If blue whales turned out to have a voiceprint—a unique acoustic signature that identified individuals—the scientists would be able to unobtrusively follow individuals in real time without having to stick a tag on them. They dreamed of conducting acoustic censuses to get more accurate population estimates than mark-recapture or line-transect surveys could provide. Many of these initial hopes have not been realized, but acoustics has nonetheless become one of the most promising branches of blue whale science. Of all the items in the researcher's tool box, including photo-ID, tagging and genetic analysis, acoustics may turn out to be the most useful for understanding the population structure of blue whales.

Before all these new recordings of blue whale vocalizations could be useful, researchers needed to build specialized instruments and learn new techniques for analyzing their data. Permanent hydrophone arrays—of which the navy's network is the largest example—can listen in on vast areas of ocean and relay the data to shore in real time, but blue whales regularly visit areas outside the range of these arrays. For decades, scientists have made recordings by dropping hydrophones over the side of a boat when they sighted whales, col-

lecting the data, then pulling up and moving on. The real promise of
bioacoustic research, however, depended on getting hydrophones
to specific areas, often in remote locations, and collecting data for
months at a time. To accomplish this, scientists devised a new kind
of acoustic recorder. There are several models, many of which are
primarily designed to record seismic activity, not whale calls, but in
general they consist of a hydrophone tethered to a plastic shell that
houses a hard drive and batteries. This is attached to a weighted
metal frame, and when the assembly is dropped from a ship, it sinks
to the bottom, while floats keep the hydrophone suspended above
the seafloor, sometimes right in the deep sound channel. Once in
place, these instruments can record low-frequency underwater
sounds for many months, often longer than a year. When the researchers return, they send a signal to the computer, telling it to cut
itself free from the frame. The assembly then floats back to the surface, where the data can be uploaded before wiping the hard drive
and deploying it again. Depending on the research group deploying them, these instruments go by different names: some are called
acoustic recording packages (ARPs), while others are nicknamed
"pop-ups."

Unfortunately, these devices don't always cooperate; hard drives
and hydrophones malfunction, batteries die, and occasionally the
release mechanism fails and the recorder is forever consigned to
Davy Jones's locker. When all goes well, however, recoverable
hydrophones can provide enormous amounts of data, and by the
late 1990s the body of literature on blue whale acoustics snowballed.
Much of the early work was done in the western North Atlantic and
the eastern North Pacific, where the blue whale populations were
particularly well studied. The vocalizations now started to be char-
acterized. Several papers, for example, described blues off the U.S.
west coast making two-part, low-frequency calls. The first part was
a rapidly repeating, amplitude-modulated pulse—that is, the call
unit was at the same frequency, but it varied in volume. The second
part began at one frequency then swept down a few hertz to a lower

tone. Researchers began referring to the monotone pulse as an "A call" and the downsweep as a "B call"—the names had been used in military documents decades before, and they would soon become part of the scientific parlance and familiar to everyone studying blue whale vocalizations.

Researchers also identified a C call, a brief upsweep that often precedes an A or B, though it is difficult to detect unless the whale is very close to the hydrophone. A fourth category, D calls, are usually the only ones audible to humans. They vary widely, though they too are downsweeps that sound rather like a tractor trailer downshifting. D calls are still extremely low-pitched, sweeping from about 90 to 30 Hz—play a recording on a good set of speakers and they will rattle the pictures on your walls—though Catherine Berchok jokingly refers to them as "their squeaky voice."

Detecting these calls is not as straightforward as dropping a hydrophone in the water and playing back the recording. Blue whales make the most energetic sounds in the sea, but because most of their vocalizations are infrasonic, humans can only hear them by playing the recordings at several times their normal speed. (Recordings of A and B calls released to the public are unavoidably sped up like this, which is disappointing, since they do not give any sense of the power the calls must exhibit to other whales.) Clearly it's impractical to listen to hundreds of hours of recordings on fast-forward. If scientists were forced to do so, most would quickly abandon the field of whale acoustics for something less excruciating. Instead, the calls are not heard, but seen: researchers find them by visually examining spectrograms, or graphs representing frequency and amplitude over time. Unfortunately, this is complicated by the fact that wind, seismic activity and the thrumming of ships are all within the frequency range of blue whale voices. Scientists learned early on that filtering out those frequencies would erase the calls as well as the noise. What they needed was software that could separate the acoustic wheat from the chaff. The good news was that certain blue whale calls do not vary much. The B call of eastern North Pacific blues, in particular, reliably descends about 2 Hz in 15 seconds or so. The call also

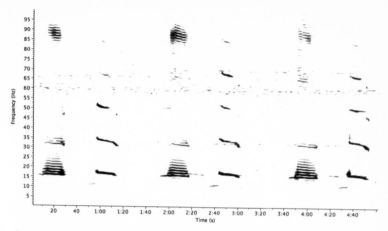

Spectrogram from an eastern North Pacific blue whale. The dark areas centred on 20 Hz are A calls; the thin parallel lines are B calls with their harmonics.

has clear harmonics, or overtones with frequencies that are multiples of the fundamental tone; for a B call centred at 15 Hz, there's a harmonic at 30 Hz, 45 Hz, 60 Hz, and so on. These form an obvious series of parallel stripes on a spectrogram, making the call easier to spot. Once they figured this out, scientists created a template with these characteristics and programmed their computers to search for patterns that matched it. Using this method, they can now mine countless hours of recordings for useful data.

As those data began to roll in, several characteristics of blue whale vocalizations emerged, some rather surprising. First, calls from the same population show remarkably little variation. With a few exceptions, all the blue whales in a given area call at the same frequencies, for the same duration and in the same patterns. Throughout the eastern North Pacific, for example, the usual calling pattern is ABABAB or ABBBABBB, with the interval between parts and between calls both remarkably consistent. In the North Atlantic, the patterns are also very regular, but they are distinctly different from those in the eastern North Pacific. Comparing the new data with the small number of older recordings from the 1960s and onward, scientists also discovered that blue whale vocalization patterns had changed only slightly over several decades. A recording of a blue whale off Chile today would show many of the same characteristics as the ones captured on tape by Cummings and Thompson in 1970. Blue whales do not have anything approaching the vocal repertoire of humpbacks, who make a whole constellation of sounds and continuously compose new songs. If humpbacks are a symphony orchestra, blue whales are a four-chord garage band.

These discoveries dampened some early hopes—identifying individual blue whales by their voiceprint, or estimating population size by the number of calls, now looked to be impossible. However, they raised another exciting possibility. If all the blue whales in a given area have distinctive call patterns, then these sound signatures, or dialects, might shed light on the animals' global population structure. The idea borrows something from anthropology; human cultures are usually delineated not by physical characteristics or geographical boundaries, but by language. Maybe the same is true for blue whales.

One of the first scientists to recognize this potential was Kate Stafford. While working on her master's at Oregon State University in 1996, she piggybacked on the work of American geophysicists who were investigating the East Pacific Rise, a seafloor spreading area that runs roughly parallel to the west coast of South America. To measure earthquakes there, the researchers moored six pop-up hydrophones on the ocean floor—two right on the equator, two more at 8°N and another pair at 8°S—and Stafford was able to get months of their data. One of the hydrophones happened to be in the vicinity of the Costa Rica Dome, and Stafford wondered whether recordings from that area could shed light on whether the whales visiting this tropical region might be the same ones seen off California.

Stafford's colleagues recovered the instruments from the seafloor about every six months and transferred the data to tape, collecting more than 47,000 hours of recordings. When she looked at the data from the hydrophone closest to the Dome, there was a clear pattern: the calls were more frequent during the northern hemisphere winter. Meanwhile, navy arrays off the U.S. Pacific

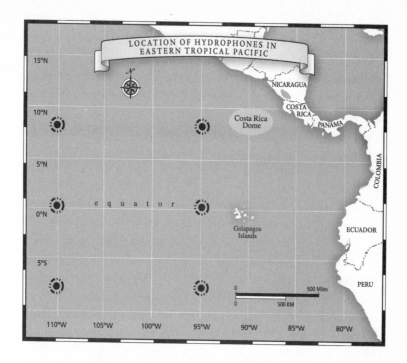

coast showed the opposite pattern—more blue whale calls in the summer—suggesting a north–south seasonal migration. In addition, the calls recorded in the two regions had the same signature. "Recordings from the vicinity of the Costa Rica Dome," Stafford wrote, "clearly indicate that the blue whales in this area belong to an eastern North Pacific stock of blue whales and not a southern hemisphere population as has been previously suggested."[10] This result was published shortly before John Calambokidis and Bruce Mate announced they had tracked a satellite-tagged whale from California to just north of the Dome, and the evidence from each method turned out to complement the another. "When I first saw the eastern North Pacific calls on the hydrophones, I got chills," Stafford remembers. "It wasn't completely unexpected, but it was really spectacular and exciting to see it, because in my mind—although, unfortunately, not in many other people's minds at the time—it meant that they had to be the same whales." Stafford found that although there was a seasonal peak—calls were most frequent from November through May—eastern North Pacific whales were heard near the Dome year-round. This suggests there is probably no resident population, as some researchers believed, but rather that some eastern North Pacific animals stay behind each year while the others migrate northward.

Another of the hydrophones revealed more secrets about blue whales in the eastern tropical Pacific, and opened a new can of worms. The instrument at the southeast corner of the array—some 900 miles off the coast of Peru—detected hundreds of hours of four-part calls that were completely different from those in the North Pacific. Yet they were almost identical to those recorded off Chile more than 25 years before by Cummings and Thompson, whose ship had been almost 2,500 miles south of Stafford's hydrophone. The 1970 recording was made in May, during the southern hemisphere autumn, while Stafford's hydrophones heard the calls most often later in the year. That suggested the population might migrate up and down the coast of South America. More surprisingly, Stafford later detected calls that were identical to those made by Antarctic blue whales, proof that at least some blues also migrate between the Southern Ocean and the eastern tropical Pacific, though these likely represent a tiny percentage of the Antarctic population. This means blue whales from two different southern hemisphere populations visit the same tropical region in winter.

Indeed, some of the blue whales mixing in the eastern tropical Pacific may inhabit both hemispheres. One of the hydrophones mounted right at the equator logged 215 hours containing eastern North Pacific calls and 535 hours with Chilean-type calls. These usually alternated with the seasons, suggesting the whales were visiting during their respective winters, but that wasn't always the case. "What this means," Stafford explains, "is that not only do blue whales from both hemispheres occur in the eastern tropical Pacific at different times of the year, but they co-occur occasionally at the equator." As if things weren't already curious enough, Stafford also detected a small number of calls she identified as *western* North Pacific vocalizations, which are more often heard off northern Japan, the Kamchatka Peninsula and the western Aleutians. If some of these blue whales are venturing into the waters off South America, they're an awfully long way from home.

Stafford's later work took her from the tropics to the Gulf of Alaska. Blue whales were once hunted there regularly, but surveys in the post-whaling era routinely failed to find them. Calambokidis and others were beginning to think the whales that once visited Alaska were part of the same population that was now turning up off California, while others argued that they had been a different stock, perhaps related to those in the western North Pacific, and had been wiped out by whaling. In October 1999, scientists dropped an array of hydrophones in the Gulf to see what they could learn, and Stafford retrieved the recordings in 2000 and 2001. Not only did she discover that there were indeed blue whales off Alaska, but she detected calls of both eastern and western North Pacific populations, occasionally at the same time, usually in autumn. As in the tropics, then, blue whales from two different populations overlap in the Gulf, something that would have been impossible to determine without acoustic data, since no photo-ID catalogue exists for the western group. Both the eastern tropical Pacific and the waters off Alaska are rich in krill, at least for part of the year, so it's possible that some whales are travelling thousands of miles from both the west and the south to converge on these feeding grounds.

If blue whales from two different populations sometimes visit the same places at the same time, it raises the question of whether there may be any mixing between the two. This would certainly be unusual; populations, by definition, are discrete groups of the same species. Often they have little opportunity to mate with one another, being separated by distance or some geographic boundary. But what if two whales did strike up a relationship in the Gulf of Alaska or the eastern tropical Pacific? The first problem with this idea is that any mixing of the populations seems to happen in feeding areas, where no mating or serious courtship is known to happen. It is also possible that blues from different areas use their distinctive vocalizations to tell each other apart and remain separate, as is the case with some birds and frogs. Stafford and her colleague Sue Moore, however, found a tantalizing piece of evidence that suggests there may be some mingling going on. In the Gulf of Alaska in October 2001, they recorded what they believe was a single blue whale emitting calls that had both eastern and western characteristics. The animal's B call resembled those of the eastern population, but with a little upsweep near the end that is regularly heard in blues around Japan and Kamchatka. Meanwhile, the A call was sometimes replaced by a typically western call. Could this be a bilingual hybrid with one parent from each side of the Pacific? "Or maybe it's an animal that is learning a song and fiddling with it," Stafford offers. "We just don't know whether it's innate that these animals produce sounds particular to their population, or whether it's something they learn."

As intriguing as this finding was, the fact remained that when it was published in 2003 no one had actually *seen* a blue whale in the Gulf of Alaska or off the Aleutians in three decades. The following summer, however, Stafford was aboard a research cruise in these waters, primarily to look and listen for humpbacks, and to the delight of the crew they photographed a blue whale off Alaska in July. Stafford may have been the only one on board who wasn't surprised. "We've been hearing eastern and western North Pacific whales on deep water recordings in the Gulf of Alaska," she told a reporter a few days later. "We knew they must be there."[11] The question of which population this whale belonged to was soon answered when the crew sent the images to John Calambokidis, who quickly identified it as one he had seen off California—the first confirmed link between the two areas. The ship then headed west, and on August 19, the crew spotted another blue whale off the central Aleutians. Four days later, they found two more. During each of these encounters, Stafford dropped a hydrophone into the water and detected only one acoustic signature: the one associated with northwest Pacific blues. After years of hearing them at these high latitudes, Stafford and her colleagues were finally able to lay eyes on whales from both North Pacific populations in the same season.

What of the North Atlantic blue whales? What can acoustics tell us about how the populations in that ocean are organized? The photo-ID work of Richard Sears, while far from conclusive, has so far pointed to two distinct populations. The western group, part of which visits the St. Lawrence every summer, may range as far north as the Davis Strait and at least as far south as Bermuda. On the other side of the ocean, meanwhile, is a presumably separate population that roams the waters from Iceland, past the Azores, at least to northwest Africa. Sears has never made a photo-ID match between these two groups, suggesting they do not cross the Mid-Atlantic Ridge. (Iceland, the Azores and Bermuda are all peaks along this undersea mountain range.) Given what Stafford and her colleagues have learned in the Pacific, then, one could reasonably predict that these two populations might have distinct vocalizations. But if there is one thing blue whale scientists have figured out, it's that nothing about these animals is predictable.

By 2003, Chris Clark and fellow Cornell scientist David Mellinger had published the results of their extensive acoustic work with the IUSS arrays in the North Atlantic. The most common blue whale call type they recorded had a two-part pattern: an 8-second pulse followed a few seconds later by an 11-second downsweep. Unfortunately, they did not separate their data by region, so if eastern and western blues make different calls, Clark and Mellinger would not have detected that. More recently, Kate Stafford and her colleague Sharon Nieukirk examined data from six acoustic recorders, three on each side of the Mid-Atlantic Ridge, each about 500 miles apart. The northernmost hydrophone was at the latitude of North Carolina, while the most southerly was at the level of Guatemala. From these locations, the array would not hear anything in the St. Lawrence or off Iceland, but if the blue whales who feed in those areas head south in the winter, they would almost certainly come within range, whatever side of the ridge they happened to be on.

When Nieukirk and Stafford collected their two years' worth of data, they heard blue whales most often on the northern hydrophones—not surprising, given that the species has never been common in tropical Atlantic waters, where the southern hydrophones were moored. The seasonal differences were also what one would expect: far more calls during the winter. Like Clark and Mellinger, they identified two phrases: one was a pulsed A call, the other a combination AB call, but there were no detectable differences between the recordings from the eastern and western hydrophones. Does this mean, Sears's findings notwithstanding, that there is just one population of blue whales in the North Atlantic, not two? Or does it mean that only the western group is calling within range of the hydrophones? Could whales from the eastern population be present but simply not vocalizing? All of these scenarios are unlikely. The simplest explanation is that both the eastern and western calls have the same frequencies and duration, even though they do not appear to overlap geographically.

Nowhere else in the world do two populations of blue whales have the same vocal repertoire, so how might this apparently unique situation have come about? Evolution is a slow process: separate a population of animals into two groups and they will eventually diverge in appearance, behaviour and other traits, but the process takes generations, centuries, even millions of years. Perhaps there was just a single population of blue whales in the North Atlantic until fairly recently. The animals that were hunted in such great numbers off northern Europe in the 19th century may have dispersed, with some heading down the European coast and others moving west toward Greenland and Canada. If they only recently took up residence on opposite sides of the ocean basin, they may simply not have had enough time to evolve unique vocal patterns. Another possible explanation is that blue whales in the North

Atlantic—which is a fraction of the size of the North Pacific—were never really out of acoustic range of one another, and therefore had no real opportunity to evolve independently.

For now, these speculations about the blue whales of the North Atlantic remain tenuous. But more than a decade and a half of acoustics research has given scientists a far better picture of the global population structure. The different song types suggest there are at least nine discrete groups.[12] We have already discussed four of them: two in the North Pacific, one off western South America, and one in the North Atlantic. There are also three distinctive call types in the Indian Ocean, another in the Antarctic and, finally, one that has only been detected off New Zealand, an area where the blue whale population is not well studied. Our understanding about the worldwide distribution of blue whales is far from complete, but at least scientists now know what to listen for.

Erin Oleson pulls the cap off a yard-long silver tube and spills the contents onto the deck of the *John H. Martin*, a 56-foot former fishing vessel now owned by Moss Landing Marine Laboratories on the shores of California's Monterey Bay. On this overcast and cool day in August 2005, Oleson roots through a spaghetti-like mass of wire, which holds an array of 13 matchbook-size hydrophones designed to detect undersea sounds of all frequencies. Oleson then pushes the whole affair overboard and watches as a carbon-dioxide canister inflates a teardrop-shaped orange balloon. The float bobs silently at the surface while everything else disappears into the sea.

Oleson, just 28 years old and a recently minted PhD, is aboard the *John Martin* to gather data for what has become the focus of her research: figuring out the behavioural context of blue whale calls. The work of Kate Stafford and others has revealed that each blue whale population has a distinctive dialect, and that individuals within a population sound more or less the same. But the *why* of blue whale calls is just beginning to be understood. Do blues call when they are feeding or travelling? Do vocalizations play a role in courtship or short-range communication? Are the whales using sound like a long-range locator beacon, as Roger Payne first proposed, or as low-frequency sonar? Are they detecting changes in sea temperatures, which may help them locate food-rich upwelling areas?

Oleson knew that before tackling any of these questions she would need acoustic data from known individuals, which she wasn't likely to get from seafloor-mounted hydrophones. She and her colleagues were going to have to get out in boats with their cameras, biopsy darts, bioacoustic tags and hydrophones. They would need to see and hear calling blue whales at the same time, something previous researchers had rarely been able to do. Only then would they be able to make the connections.

The first step in this quest was to learn whether one or both sexes vocalize. If only female or only male blues emit calls, they surmised, the animals probably are not using the technique for navigation or feeding, since both sexes need to eat and find their way. A gender difference was more likely to indicate a reproductive role. Scientists have long known that only male humpbacks sing, for example, and the usual theories are that they use songs to attract females or to challenge other males, or perhaps both. Perhaps blue whales were doing something similar. The earliest attempt to determine if gender played a role in calling came during a 1997 research cruise off California that included John Calambokidis. On their most productive day, the scientists heard calls from five blue whales in the Santa Barbara Channel, three producing A and B calls and two making D calls. The team had five hydrophones in the water, making it possible for them to track the whales' movements quite accurately, although only two of the animals were actually seen. Calambokidis managed to biopsy one of the whales emitting A and B calls, and it turned out to be male. The scientists later biopsied two other whales, neither of which appeared to be calling. These silent swimmers were both female. It was a good start, but even by the time Erin Oleson's graduate research turned to blue whales, no one knew for sure whether all calling blue whales were male. Which takes us back to the *John Martin* in Monterey Bay.

Since 2000, Oleson and her colleagues have been making regular excursions, both day trips and longer cruises well offshore, to record calling blue whales, to observe their behaviour simultaneously, and to try to match the callers with the individuals in Calambokidis's catalogue. The instrument she's unpacking this August day is a DIFAR sonobuoy (for "direction finding and ranging"). It includes 13 vertically spaced hydrophones, a magnetic compass and directional sensors. The hydrophones used in bottom-mounted recorders or dropped overboard from ships are usually omnidirectional, which means they can pick up sounds from all directions, but they cannot tell the listener where those sounds are coming from. DIFAR sonobuoys are different: they can accurately measure the bearing of a calling whale and, when two or more are used in conjunction, they can also give an idea of its distance. The transmitter then sends all of this data by VHF to Oleson's computer in real time. Oleson and other researchers scrounge these sonobuoys from the navy after they've exceeded their shelf life. They were originally designed to be dropped from aircraft and used to detect submarines, but like the IUSS arrays, they have found a new usefulness in the field of whale acoustics.

Once her sonobuoy is in the water, Oleson moves into the cabin of the *John Martin* and opens her laptop, where she can instantly view the spectrograms of the sounds detected by the hydrophones. It isn't long before a calling blue whale, nowhere to be seen on the water, appears on her screen. Dashed horizontal lines indicate the monotone pulses of A calls, while a vertical stack of four or more parallel lines, all curving downward from left to right, is the obvious signature of a downsweeping B call and its harmonics. When Oleson fixes the approximate location of the calling whale, she gets on the radio to Calambokidis, who is patrolling nearby in his inflatable. "If I can figure out, for example, that there's a whale to the north, I can send John in that direction," Oleson explains. "Hopefully he can find the whale that was making the sounds, and he can tag that whale." He'll also get a photo of the animal and, with some luck, a biopsy. The tag Calambokidis will try to attach is the same bioacoustic probe that has revealed so much about blue whales' diving behaviour. Indeed, while the B-probe contains a suite of instruments unrelated to acoustics, it was originally designed to detect vocalizations and record them on an internal flash drive. By analyzing these recordings and the data from the tag's other sensors, Oleson will be able to determine the precise depth of the calling whale, and whether it's travelling in a straight line or making movements associated with feeding. She also wants to know whether the whale is calling alone or in the presence of others. Once Calambokidis deploys the B-probe, the crew of the *John Martin* will track the animal with a VHF receiver and pick up the tag when it finally falls off.

Even after several field seasons, Oleson and her colleagues are a long way from decoding the language of blue whales, but they have made some fascinating links between calling and behaviour. To begin with, they are now quite sure that only males make A and B calls. In one study, they got biopsies of eight known and two suspected AB callers, and all were male. They also took samples from 17 non-calling whales, only three of which were male. However, they also biopsied three whales who were making D calls, and one of these was female. So both sexes do vocalize, but with different repertoires.

Oleson also found a distinction that others hadn't noticed before. Blue whales in the eastern North Pacific had long been known to make A calls followed by one or more B calls, and to repeat these patterns over and over in "songs." But Oleson found A calls that were *not* followed by a B, as well as B calls not preceded by an A, and these intermittent calls showed no regular, songlike pattern. As it turned out, whether the whales were singing long songs or making singular calls depended on what they were doing. The singers were always alone, and always travelling, moving in a more or less straight line at 4 to 9 knots, without making feeding dives. The singular callers, on the other hand, could be travelling, feeding, resting or making short dives. More interestingly, they were in pairs or small groups, and when Oleson's team was able to collect tissue samples from these associated whales, the male caller's partner was always female.

As for the whales making D calls, she found a pattern there, too. The calls occur between feeding lunges—the animals usually dive 250 feet or more into a krill patch, come back to the surface to breathe, and then make another shallower dive before emitting the D call. Two of the three D callers the scientists tagged were in loosely associated pairs; that is, close to one another, but not synchronizing their diving or breathing. In one case, the tag recorded calls at quite different volumes, suggesting that both whales in the pair were vocalizing. The louder calls probably came from the whale carrying the tag, while the softer ones were picked up from a whale swimming and calling some distance away.

The depth sensors and accelerometers on the tags added more pieces to the puzzle: the whales made all three call types less than 100 feet below the surface, and they were generally moving horizontally, not diving or ascending. Although blues can be surprisingly agile, often arcing, rolling and even inverting their enormous bodies while feeding, they were oriented almost perfectly upright when vocalizing. One reason that blues are calling at these shallow depths is probably physiological; although they are known to dive up to 1,000 feet when feeding, the pressure at even a third of that depth would likely compress their bodies so tightly that producing a call would be impossible. Oleson says there are other advantages to calling at 65 to 100 feet. For one thing, this is approximately the depth at which a blue whale is neutrally buoyant, so the whale can focus its energy on vocalizing without needing to actively swim. "The other thing is, if they're sitting at 65 feet or so, the sound will bounce off the surface and then come back down, so the call will be louder in the downward direction than it would be in the horizontal plane." Unable to point their calls the way dolphins do, blue whales may be steering them into the deep sound channel by indirect means. "We have no evidence that they're doing that," admits Oleson. "At this point, it's just conjecture."

So what can we make of all this? First, if repetitive AB songs are made only by solitary, travelling males, it seems likely that the whales are engaging in some kind of long-range communication, probably with a connection to reproduction. Males may be advertising their presence in an effort to attract females, similar to what humpback and fin whales do, even during the feeding season. To add support to this interpretation, the males sing more often later in the year. Sears and Calambokidis have both found that blues are more likely to form long-term pairs toward the end of the summer and into fall, so perhaps the increased singing is another indication that males are actively looking for a partner as winter approaches.

Oleson also offers an explanation for why the singing whales are always on the move. Because blue whales have such extreme energy demands, she suggests, they may need to focus all their attention on feeding when they happen upon a swarm of krill—they can't afford to be singing when they should be eating. Instead, the animals seem to save their longest vocal performances for the times when they are travelling between feeding areas, like an executive using her hour-long commute to return the phone calls she missed during lunch.

The singular AB calls are harder to figure out, since they occur during many activities, from feeding to travelling and everything in between. However, because the calls are made by males in pairs or small groups, Oleson feels that these nearby whales are likely the intended audience. Unlike the travelling singers, who may be broadcasting their presence far and wide in search of females, males making singular AB calls may be trying to maintain pair bonds they've already established by saying something along the lines of, "Don't go away, I'm still here."

As for the function of the D calls, things are even less clear. Because both sexes make these vocalizations, and because the calls do not seem to vary much between different blue whale populations,

Oleson believes they are related to social interaction rather than reproduction. But how? If blues are making these calls during breaks from feeding, could they be trying to attract others to the area? Since blue whales don't seem to feed cooperatively, this seems unlikely. "Maybe there is some premium on maintaining contact with a specific individual," Oleson ventures. "Even if you're not cooperatively feeding, you may want to keep in contact with a specific whale, whether that be a mother and calf, or a pair that has developed and wants to keep track of each other." She points out that other baleen whales keep in acoustic contact by using calls with a distinctive frequency sweep: right whales make an "up-call" that gets higher, while finbacks make a downswept call similar to that of blue whales. These little changes in frequency make the calls more detectable amid the background noise, and the whales can more easily tell which direction they're coming from. Both right and fin whales have been observed approaching vocalizing whales of their own species, suggesting that the calls are used to keep groups together.

Oleson's other research interests are equally promising. Along with Calambokidis and others, she is exploring ways of using both sightings and acoustic detections in tandem when making population estimates. After all, sometimes you hear blue whales you don't see, while other times you see whales you don't hear, so combining the results of both survey types should improve the results. As for some of those other early goals of acoustics researchers, such as identifying individual blue whales or tracking them in real time? "Honestly, I don't think we're any closer to that," Stafford says. "I thought we would be."

NINE

THE 79-FOOT PYGMY AND OTHER STORIES

WHILE THE BLUE WHALES of the North Atlantic and North Pacific have distinctive call types, they have no significant morphological differences. A scientist examining a specimen would not be able to tell you whether the animal came from the Gulf of St. Lawrence or the Santa Barbara Channel, at least not without analyzing it at the molecular level. That is not the case in the southern hemisphere, however, where there are at least two subspecies with distinctive body sizes and shapes: the Antarctic blue whale and the smaller pygmy blue.

Antarctic blue whales—the largest of all the world's populations—may once have included individuals over 100 feet long. The Southern Ocean used to teem with these giants, but they were rarely encountered north of 55°S in summer. Pygmy blues, on the other hand, were first identified in the southern Indian Ocean, well north of Antarctic waters, and of the thousands killed by Japanese and Soviet whalers, none was ever measured at more than 79 feet. That's a good 20 percent shorter than the largest of the Southern Ocean blue whales, though it is still rivalled in the animal kingdom only by the fin whale. The term "pygmy blue whale," indeed, is something of an oxymoron.

The inappropriate name is probably part of the reason why the classification has been controversial. Many so-called pygmy species are radically different from their larger namesakes—the pygmy sperm whale is no more than 10 feet long, and the pygmy right whale is usually less than 20 feet. Neither is in the same taxonomic family as the great whales. Pygmy blues, in sharp contrast, are indistinguishable from "true blues" at first glance. Telling the two subspecies apart from aboard a ship has long been a problem, because they can overlap both geographically and in size: a 75-foot blue whale near the Antarctic convergence might belong to either subspecies. This has been a glaring problem for people trying to determine whether Antarctic blues are increasing after their decimation during the whaling era. If their ranges really do overlap and you cannot distinguish between them, then getting an accurate count of each subspecies is impossible. It would be like walking through London and trying to count Englishmen while excluding people who look Welsh.

To confuse things further, southern blue whales outside the Antarctic traditionally have been lumped together as pygmy blues, but recent research suggests that not all of the populations are closely related. The blue whales that feed off Australia and Indonesia have long been accepted as pygmies, but things are much less clear in the southeast Pacific, where blues congregate off Chile in the summer. Meanwhile, a little-studied population off Sri Lanka, the Maldives and around the Arabian Peninsula—in the low latitudes of the northern hemisphere—is sometimes singled out as a fourth subspecies, but now appears to be indistinguishable from the pygmy blues on the other side of the equator. Bottom line, there is serious taxonomic turmoil surrounding blue whales, just one more indication that humans still do not understand some fundamental things about these animals.

Blue whale taxonomy, as it happens, has been plagued with confusion since the 18th century. The Swedish botanist Carolus Linnaeus got everyone off on the wrong foot in 1735 when he devised his first system for classifying living things and included whales as a variety of fish. It wasn't until the tenth edition of his *Systema Naturae*, 23 years later, that he identified them as mammals. In that edition, Linnaeus assigned a name to the 78-foot specimen that had stranded in 1692 in the Firth of Forth, where it was described by Sir Robert Sibbald. But the name he gave this giant whale, *Balaena musculus*, has left historians scratching their heads ever since. *Balaena* is clear enough, but *musculus* is Latin for "little mouse." (The common house mouse is *Mus musculus*.) What on earth was Linnaeus thinking in giving this moniker to a 78-foot sea creature?

There are a few possible explanations, though none is terribly satisfying. The Latin *musculus* may be related to the French and English words for "muscle," so perhaps Linnaeus was referring to the animal's enormous bulk. Others point out that Pliny the Elder, the 1st-century Roman naturalist, used *musculus* to refer to a fish having "no teeth at all, but in place of them, the interior of the mouth is lined with bristles,"[1] which sounds a lot like a baleen whale. (Here he is borrowing from Aristotle, who four centuries earlier wrote that "the so-called mousewhale instead of teeth has hairs in its mouth resembling pigs' bristles."[2]) Pliny's reference is ambiguous, though, because elsewhere he uses the same name to describe a fish that guided whales through the sea, "for, as the eye-brows of the [whale] are very heavy, they sometimes fall over its eyes and quite close them."[3] It is not clear whether *musculus* is the fish with the bristles or the one that guides blind whales. George Small, for his part, believed that "the only plausible explanation for the choice of the term is that Linnaeus must have been in a jocular mood at the time."[4] The father of taxonomy did have a flair for language and a mischievous streak: once, when a fellow botanist, Johann Siegesbeck, harshly attacked his method of classifying plants, Linnaeus named a weed after him. So maybe he really was indulging in an inside joke.

Linnaeus named four whale species in his 1758 edition. His *Balaena mysticetus* was the bowhead, and we now know that his *B. physalus* was a fin whale, while his *B. musculus* was a blue. (The fourth, it was later determined, was also a fin whale that Linnaeus had mistakenly identified as a new species.) But not surprisingly, given their limited opportunities to observe large rorquals alive or dead, scientists over the next 150 years made a mess of classifying the blue whale. In 1804, a French naturalist coined a new genus name, *Balaenoptera*, combining the Latin for "whale" with the Greek *pteron* ("wing"), a reference to the dorsal fin common to rorquals but absent in bowhead and right whales. This was a step in the right direction, and the new name was adopted by the English explorer and scientist William Scoresby when he published his important work, *An Account of the Arctic Regions with a History and Description of the Northern Whale-Fishery*, in 1820. However, Scoresby got his fin whales and blue whales mixed up: he described *Balaenoptera physalus* as "the longest animal of the whale tribe; and, probably, the most powerful and bulky of created beings," and wrote that *B. musculus* is shorter and "is said to principally feed on herrings."5 In 1846, another English taxonomist muddied the waters again when he classified the largest whale of the North Atlantic as *Balaenoptera sibbaldii*.

Over the next few decades, scientists in other parts of the world set about giving new names to the giant whales of other ocean basins. The English zoologist Edward Blyth, a colleague of Charles Darwin, settled in India in 1831 and took charge of the collections of the Royal Asiatic Society of Bengal. In 1859, Blyth published a paper in that society's journal describing two gigantic rorqual specimens: one, reported to be 90 feet long, had washed up in Bangladesh, while the other, purportedly 84 feet, was found on the shores of a Burmese island. Blyth examined some of the bones of the latter animal and pronounced it a new species, "which I think we can safely venture on designating *Balaenoptera indica*."6

Thirteen years later, in 1872, the German biologist Hermann Burmeister, working in Argentina, described three rorqual specimens from the South Atlantic. One was 58 feet long and "rather young," but "as I did not see the animal before the body had been dried, I cannot accurately describe its external appearance."7 However, Burmeister notes the whale's baleen was entirely black, almost certainly identifying it as a blue whale, since the baleen of the other rorquals is lighter in colour. He declared the animal a new species and named it *Balaenoptera intermedia*.

Finally, in the United States, the paleontologist Edward Drinker Cope, perhaps best remembered for his work on dinosaurs, coined his own scientific name for the largest whales found off the Pacific coast of North America. Combining the name of Sir Robert Sibbald and the still common term "sulphur-bottom," Cope dubbed the animal *Sibbaldius sulfureus*. In 1874, when the former whaler Charles Scammon published his highly influential book, *The Marine Mammals of the North-western Coast of North America*—the one in which he described sulphur-bottoms being blown up with bomb-lances off California—that is the name he used.

It is worth pausing here to recall that 19th-century scientists had little or no opportunity to compare their whale specimens with those examined by their colleagues, especially those on other continents. As a result, it was quite common to identify a "new species" that had already been described and classified by another scientist, and that is exactly what happened with blue whales during this period. Had Blyth, Burmeister and Cope been able to compare their specimens with the whales Svend Føyn was beginning to haul out of the sea in Europe, they would have realized that they were all the same animal. *Balaenoptera sibbaldii*, *B. indica*, *B. intermedia*, *Sibbaldius sulfureus*, not to mention several other short-lived monikers, were simply blue whales by different names.

Science needed someone to clean up this muddle, and that person turned out to be Frederick True of the United States National Museum, now part of the Smithsonian Institution. True, the museum's first head curator of biology was a specialist in cetaceans. He carefully worked his way back through Linnaeus's sources and found a number of errors in the Swede's work. He also discovered that later scientists often used the name *musculus* to refer to fin whales, not blues, as Scoresby had in 1820. (Part of the confusion may have come from the practice of referring to all rorquals as "finbacks.") In 1899, True published a paper suggesting that *sibbaldii* be dropped, and that Linnaeus's species names be restored to the correct animals: *physalus* for the fin whale, and *musculus* for the blue. As for *indica* and *intermedia*, large whales in the northern Indian Ocean and off Argentina were so rarely encountered at that time that no one much worried whether those names were valid. By and large, other scientists followed True's advice, and as the era of pelagic whaling dawned, all blue whales in both hemispheres were generally referred to as *Balaenoptera musculus*. (As late as the 1950s, however, a few scientists still preferred the genus name *Sibbaldius*.)

It was not until the dying days of blue whale hunting that the taxonomy became contentious again. Blue whales were by then so scarce that only a few hundred were being taken each year, and many of the IWC's scientists were suggesting a complete ban on taking blues in the southern hemisphere. The hard-liners resisted, however, and during this debate the Japanese caught a lucky break. In March 1960, they killed 311 blue whales in a newly discovered whaling area in the south-central Indian Ocean, near the Kerguelen Islands, which are about equidistant from southern Africa, Australia and Antarctica. In 1961, the biologist Tadayoshi Ichihara published an article describing these whales as a new population, distinct from those that summered in the Southern Ocean.[8] Not only were they geographically separate, they were also physically different: considerably smaller, with a less elongated body shape. Ichihara soon began calling this animal the pygmy blue whale and later gave it the Latin name *Balaenoptera musculus brevicauda*, arguing that it qualified as a distinct subspecies on a number of grounds. First, their summer feeding habitat was primarily outside the Antarctic, north of 54°S, and they preyed on a different species of krill. They also had "a much shorter tail and a noticeably longer trunk" (*brevicauda* means "short-tailed"), a more silvery grey skin colour, and shorter baleen plates "suggesting a difference in skull shape," although Ichihara admitted he did not actually have any skull measurements to go on. Most of these whales were 67 to 72 feet when fully grown, and the largest female was 79 feet, about one-fifth shorter than the biggest Antarctic blues.

A number of scientists immediately disputed the whole idea of a pygmy blue whale. At the conference in Washington, D.C., where Ichihara first presented his findings in 1963, delegate F.C. Fraser announced that he had examined many blue whales in South Georgia during the 1920s, and that many were quite small. The Norwegians even had a name for these runts: *Myrbjønner*. "We did not recognize these as being different from the ordinary blue whales," Fraser explained to Ichihara. "They were just small, mature blue whales. I wonder if there has been any examination made of these *Myrbjønner* in relation to your pygmy blue whale." Ichihara replied that the small blues captured off South Georgia were pygmies that had ventured into Antarctic waters. Unconvinced, Fraser persisted: "The trouble is that there is no research as far as I know on the *Myrbjønner*." He pointed out that the catch data made no distinction between immature whales and adults that were smaller than average, so it would be very difficult to compare the *Myrbjønner* to this supposed pygmy blue. "But in my opinion," Fraser said, "it is the same whale."[9] Other biologists argued that since the IWC did not allow blue whales under 70 feet to be killed, smaller individuals would likely increase in number as they alone survived to pass on their genes. In their opinion, too, there simply was no compelling evidence for creating a new subspecies.

While Fraser and others quibbled with Ichihara's data, the budding environmental movement was far more harsh. When George Small's *The Blue Whale* appeared in 1971, the author argued that Ichihara had made up the whole business in an effort to keep at least some southern hemisphere blues unprotected. "The obvious willingness of the Japanese to kill the last blue whale was so blatantly rapacious that even the Japanese themselves felt obligated to proffer an explanation," he fumed. "In my opinion, the pygmy blue whale was a fraud used as an excuse to continue killing blue whales."[10] The conservationist Farley Mowat took a similar view, writing that the Japanese had evaded the IWC's proposed ban by "pretending to discover a new species of whale."[11]

More than four decades later, we can consider this question with less emotion. There is no doubt that Ichihara and the Japanese used their discovery as a pretext for vetoing the ban on killing blues throughout the hemisphere. At the 1963 IWC conference, they agreed to stop taking blue whales in the Antarctic—since there were almost none left anyway—as long as the area of the Indian Ocean inhabited by this "nearly virgin stock" remained open. They did so again in 1964 before finally agreeing to a total ban in the southern hemisphere the following year. But whatever ulterior motives the Japanese may have had, the consensus of scientists today is that pygmy blue whales really are a distinct subspecies. Although there is no universally accepted definition of *subspecies*, the term typically refers to a geographically separate population or group of populations with identifiable characteristics that set it apart from other members of the same species. By that definition, Ichihara's pygmy blues qualify: they inhabit a different ocean from the Antarctic blues, and they have a different morphology. The case was strengthened when the Soviets' illegal whaling came to light in the 1990s. During the 1960s and 1970s, the Soviets took thousands of blue whales in the Indian Ocean, and their catch data indicate that these animals were of similar size and appearance to those Ichihara described. The Japanese certainly exploited their data, but they did not invent it.

Once you move outside the Indian Ocean, however, the whole pygmy issue is much less clear. In fact, if you look at the global population of blue whales, pygmy blues in the Indian Ocean are not even unusually small; their average length is not much different from blues in the northern hemisphere. Jim Gilpatrick, who spent a decade making photogrammetric measurements of living blue whales and comparing them with whaling data, found remarkable similarities. "The whales we're seeing off California are similar in size to the pygmy blue whale described by Ichihara," he says. "In terms of their rostral length, their tail length, and their total length they're exactly the same. I'm not going to say they *are* pygmy blue whales, but certainly for the three external morphology characteristics we looked at, they're similar." If anything, it is Antarctic blue whales that are anomalous because they are so large.

Richard Sears has never showed much patience for this debate over what he half-jokingly calls "midget whales." He remembers having a discussion with the Japanese scientist Masaharu Nishiwaki at a marine mammal conference in 1983. The two bumped into one another in an elevator, and Nishiwaki's wife acted as an interpreter. "He was going on and on about how the blue whales in the St. Lawrence and in Baja were pygmy blue whales because of this and that phenotype." (A phenotype is a visible characteristic of an animal, such as shape or colour.) Sears bristles at this reasoning, not because he denies that blues in the Indian Ocean, or in the northern hemisphere for that matter, are smaller or less streamlined than Antarctic blues, but because he does not believe that is enough to reclassify them. "If I'm wrong, fine, burn me at the stake. But I've read the papers about pygmy blue whales and they are based on phenotypes. I used to work on salmon, and Atlantic salmon—all of which are the same species—are different from one river to the next. The river environment chisels the fish: some of them are more fusiform [tapered at both ends], some of them might be shorter, some might be bigger because they have more food. So it seems to me that the scientists studying fish aren't playing by the same rules as those who study whales." Even those who study whales seem to be inconsistent here: humpbacks from various parts of the world differ markedly in appearance, but they are all classified as the same species.

Dale Rice's widely cited *Marine Mammals of the World: Systematics and Distribution* divides the world population of blues into three diagnosable subspecies, and discusses a fourth more tentatively. All those in the North Atlantic and North Pacific are lumped together as *Balaenoptera musculus musculus*, sometimes called northern blue whales. Antarctic blue whales—occasionally called "true blues," though many scientists dislike the term—are classified as *B. m. intermedia*, using the name Burmeister coined in 1872. Rice writes that the pygmy blue, *B. m. brevicauda*, is concentrated in the south-central Indian Ocean, but goes on to say they may range as far west as the South Atlantic and as far east as the Tasman Sea between Australia and New Zealand. He also says the whales off the Pacific coast of South America may belong to this subspecies, too. Finally, he recognizes Edward Blyth's "great Indian rorqual," *B. m. indica*, as the subspecies observed off Sri Lanka, the Maldives and elsewhere in the northern Indian Ocean, though "its distinguishing features, if any, remain poorly known."[12]

This breakdown of blue whale subspecies, even when one sets aside the almost complete lack of data about *indica*, is rife with problems. The most glaring is that the animals Rice describes as pygmy blues range across three different ocean basins: the Indian, the South Atlantic and the eastern Pacific. So what began as a name for a geographically isolated population has since become a convenient but ill-defined label slapped onto any southern hemisphere blue whale found outside the Antarctic, not to mention that it has been tossed around in reference to northern hemisphere animals, too. This imprecision is not just an academic dispute with no real-world significance; it is a serious obstacle to tracking the recovery of blue whales. How can you monitor a population if you do not know where it roams and who it breeds with?

Scientists started tackling this problem in the mid-1990s, and while the work continues, they have pieced together a much better picture of blue whale populations in the south. Surprisingly, the good guys in this area turned out to be the International Whaling Commission and the Japanese government, who have funded much valuable research in the past couple of decades. One of the tasks of the IWC's Scientific Committee is continuously assessing the stocks of baleen whales and monitoring their recovery. Much of the effort since the 1970s has focused on minkes, since the Japanese still kill hundreds of these whales annually. However, members of the committee met in Tokyo in 1995 to plan a series of research cruises that would focus on southern hemisphere blue whales. These eventually became part of the Southern Ocean Whale and

Ecosystem Research (SOWER) program, and their shipboard surveys have since covered thousands of square miles in the Antarctic, the southern Indian Ocean and the southeast Pacific. The data gathered on these cruises, as well as the work of independent researchers, have compelled scientists to rethink almost everything they thought they knew about blue whales that roam the bottom half of the globe.

Peter Gill began working with humpback whales in 1983, but it was a research cruise sponsored by the IWC and Japan a dozen years later that changed the direction of his life's work. "That's when they found quite a gaggle of blues in this region," Gill says from his home near Portland, Australia, about 200 miles west of Melbourne. "That was really the first indication that this was an aggregation area for blues. Prior to that there were just incidental sightings."

Australia had a long and bloody whaling history, but the prime targets were right whales, sperm whales and humpbacks. There were occasional reports of blue whales, however, either at sea or stranded onshore, and a few dozen were caught near Perth. Some of these were almost certainly Antarctic blues that had migrated north in winter; one killed at Twofold Bay in 1908 was reportedly 97 feet. When Soviet whalers were illegally killing everything they could in the late 1960s and early 1970s, they took mostly pygmy blues off southern and western Australia. But sightings were rare in the decades that followed, which is why no one was prepared for what the research cruise turned up after leaving Fremantle in December 1995. Two vessels, both provided by the Japanese government, surveyed the waters off southwestern Australia and then travelled east, ending up in Hobart, Tasmania. Along the way they encountered a gathering of blue whales in Discovery Bay, midway between Adelaide and Melbourne. The Japanese observers on board—all experienced at visually identifying whales at sea—proclaimed the vast majority of them pygmy blues. The scientists found swarms of krill near the surface here, too, as well as the

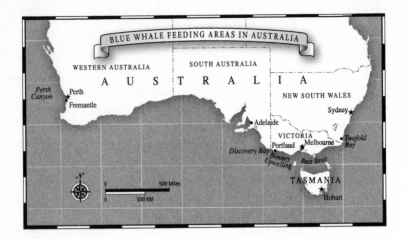

distinctively reddish feces that confirmed this was a feeding ground. Like the blues discovered off California a decade before, here was a hangout for the world's largest animals not far from an urban shore, yet no one had known it existed.

That's where Peter Gill enters the story. In 1998, he moved from the Blue Mountains, west of Sydney, and settled outside of Portland. This town of about 9,000 is a short drive from Discovery Bay, and Gill has since devoted his career to studying the blue whales in his adopted backyard. Two years later he was joined by fellow biologist Margie Morrice, who specializes in the feeding ecology of blues, and the two have been collaborating ever since, expanding their study area west past Adelaide and east to the Bass Strait between Tasmania and the mainland.

Oceanographers had already pieced together the events that conspire to attract baleen whales to these waters. This part of southern Australia is an extensive coastal upwelling zone, where summer winds bring cold, nutrient-rich water to the surface, giving rise to blooms of phytoplankton, which in turn nourish masses of krill. "The krill seems to just bubble up like a pot of soup," Gill says. "You get a hot spot here and the whales somehow find it and exploit it, and then it dies down and they'll move off." The area has since become known as the Bonney Upwelling and is now identified as one of the most important blue whale feeding areas in the world.

While Gill and Morrice also work from a small boat, they do much of their surveying from the air. "We fly our aerial surveys at 1,500 feet, because blue whales seem quite sensitive to aircraft at 1,000 feet, and we learn more about them if we're disturbing them less," Gill says. "We'll find whales and circle for 5 or 10 minutes and just see what they're doing, see how they feed. You can get some fantastic observations from the air." The topography and currents in the Bonney Upwelling nudge the krill toward the surface, so blue whales are regularly seen foraging in full view. "A very common feeding behaviour here is surface lunging," Gill explains, "where the whale rolls on its side—almost invariably onto its right-hand side. That gives them more lateral flexibility for lunging into the krill swarms. Once they roll onto their side, the spinal flexibility is sideways rather than up and down, so they can do tighter turns. You see the left pectoral fin and the left fluke come out of the water, and then they open this huge cavern of a mouth as they lunge into a swarm and fill it with ocean and krill. They become what we call the blue tadpole."

These surface-feeding blue whales allow Gill and Morrice to see parts of the animals' lives that are often hidden. "At times we've seen two whales feeding synchronously, side by side, rolling at the same moment and ploughing into the same krill swarm," says Gill. "Other times you'll see two whales hanging around an area, and one of them will feed and the other won't. But these are just short observations, where you're there for 5 minutes. You've got no idea whether they're together simply because there's a lot of krill, or whether they're pals that have been hanging around for weeks or months, or even years."

While aerial encounters are often fleeting, observing blue whales from above gives a perspective that is far different from what you get in a small boat. When you're down at sea level, it is next to impossible to determine whether that blue whale beside you is associating with a second one a couple of hundred yards away. From a vantage point 1,500 feet above, however, a relationship becomes more obvious.

Morrice remembers watching what she believes was an adult and a juvenile foraging together. "I definitely think it was cooperative behaviour—the juvenile would approach a large krill swarm and appear to wait there, and then the larger animal came in and fed on it. Then the smaller animal would go off to another krill swarm, like it was scouting—it was quite amazing." The conventional wisdom is that blues do not cooperatively feed, at least not like humpbacks, who often work together to herd fish. What Morrice's observation suggests, though, is that interacting blue whales do not need to be cheek by jowl. "You may not have animals very close together in what you'd classically call a group or a pod—they may be hundreds of metres apart, or even kilometres apart, but they're still perhaps associating and coordinating their movements."

It is too early for Gill and Morrice to get any useful estimate of the size of this population, and their study area is likely a small part of the animals' range. "We don't think that the whales we see are only coming *here* to feed," Gill says. Where else might they be going? About 1,200 miles to the west of the Bonney Upwelling, off the southwest corner of Australia, there is a food-rich area in the Perth Canyon that is also visited regularly by blues. "Our feeling is that there is interchange between these areas. We're just developing our photo-ID capability, and we've got one really good resight between here and Perth, and the calls are identical between the two areas. It's not far for a blue whale—a few days' travel. We think we're probably going to find more and more interchange throughout this Australian blue whale region." Gill and Morrice have also learned (from a satellite-tagged whale in 2004) that some of their whales also go south to the subtropical front, the edge of the Southern Ocean, where Soviet whalers illegally killed large numbers of blues. Along this front, which varies between about 39°S and 47°S, the warmer water from the north meets the colder water from the south, creating conditions that can lead to abundant plankton swarms. The distances between these areas are enormous for a couple of biologists in a boat or small plane, but for an animal with 15-foot tail flukes they're a short jaunt. "Blue whales are greyhounds," Gill says.

Peter Gill and Margie Morrice were among the first scientists on the scene after an unknown blue whale population turned up off Australia in the mid-1990s.

"We've done surveys where we've seen 30 or 40 whales in an area, and then you go back a week later and they're not there. They've moved a hundred miles or more down the coast. They've just got this incredible ability to head off looking for food somewhere else—goodness knows how they find it."

Blue whales usually show up in the Bonney Upwelling in November and are gone by late May, though where they go in winter is as much a mystery here as in most other parts of the world. One theory is that the whales move up the western Australian coast and into the tropical waters of the Indonesian archipelago. This area is unique on the planet—it is the only equatorial region that is a gateway between two oceans. More than 30 whale and dolphin species are known to inhabit Indonesian seas, and some may move from the Indian Ocean into the Pacific, and vice versa, through the islands' narrow passageways. Unfortunately, in a nation that is extremely poor, difficult to travel around and teeming with wildlife that is a whole lot easier to study than whales, knowledge about cetaceans here is rather sparse. One of the people looking to change that is Benjamin Kahn, a Dutch-born scientist who now lives in Australia and has worked the waters around Indonesia, Papua New Guinea and farther afield since the early 1990s. At the start of his career, Kahn was tracking sperm whales around northern Sulawesi and realized there were other whales and dolphins in the area that no whaler or scientist had ever noted. "Almost every day we were seeing 10 or 12 species without even looking for them, so very quickly we learned that we would have to broaden our interests," Kahn says. One of his main interests since that time has been the archipelago's population of blue whales, which are believed to be pygmies.

Scientists studying blue whales off California, in the Gulf of St. Lawrence and southern Australia can get out for a full day's work and still be home for supper. Not so in Kahn's study area. Getting from Bali to the island of Alor—across the 600-mile length of the Lesser Sundas—takes three days, and the sail from Bali to Irian Jaya (the western half of New Guinea) takes at least a week. Some of the places where Kahn follows whales are so remote that he has to charter a local vessel and crew for 10 to 15 days at a time, sailing to areas where you cannot find diesel fuel, and where ports are so limited that it's hard even to get fresh food and water. There are few anchorages because of the big ocean swells, and intense currents make navigating tricky.

When Kahn began his research on blues, simply deciding where to look for blue whales was one of his biggest challenges. "You have no idea where to start because you have 17,000 islands," he says. To narrow the field, Kahn looked at the data on historical sightings and overlaid them with what oceanographers knew about currents and upwelling zones. He concluded that his best bet was to look for places where the animals would be forced to squeeze through narrow straits between islands. "These bottlenecks concentrate large marine life—the animals have to go through them to get to the upwelling areas, so you have a much greater likelihood of encountering them. We call them sites of no regret." Kahn and his colleagues used this method to find several important corridors, including one between Sulawesi and the Philippines and a few around Papua New Guinea, all of which are gateways to the Pacific. However, the most

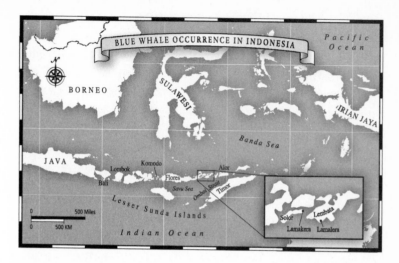

important areas for blue whales are farther south in the Lesser Sunda Islands, including a bottleneck between Bali and Lombok and an even more important one in the Ombai Strait, which separates the islands of Alor and Timor. "That's definitely the hot spot for blue whales in Indonesia, and maybe all of Asia."

Not surprisingly, the southern Sundas are home to an important upwelling zone. "In the northern Indian Ocean, there's nothing else like it," Kahn says. "It's a very interesting feature that starts around April when the monsoons pick up, and dies down again in November. Whether the area is a destination for blue whales by itself or a pit stop—a bit of fast food along the way—I'm not sure. We're slowly trying to piece that together, but it's probably a bit of both." Blues are regularly spotted in the Banda Sea, north of the Ombai Strait, suggesting that some may be passing through this corridor from the Indian Ocean, perhaps via Australia. But they will also linger here to feed. "We see animals that spend a day in a particular area, which is very different from the behaviour of other animals that are migrating through."

The fact that the upwelling peaks from April to November—the exact opposite of the situation in southern Australia—seems to suggest that blue whales move between the two areas according to the seasons. The problem, however, is that the whales don't disappear from Indonesia in summer. They are seen in every season. This is quite a different scenario from the blues off Australia, who are virtually unknown between May and November. "The whales here do seem to have a more limited range," Kahn says. "Whether it's as limited as hanging out in Indonesia all year round, I'm not sure. Maybe there is some sort of conveyor belt scenario, where we get certain individuals from the same population visiting the Indonesian seas with overlap, which makes for the year-round sighting pattern."

In 2005, Kahn decided to attach satellite tags to a couple of whales in the Savu Sea to get some idea of where they might be travelling. The tags used by Bruce Mate and others rely on satellites in polar orbits, which rarely pass overhead when you are close to the equator, making real-time transmission less practical. So Kahn's team instead used pop-up archival tags, which store time and dive-depth information on a hard drive. The tags are then programmed to release themselves from the whale's back after a certain number of days and transmit their stored data via satellite, saving the researchers from having to find and recover the tag in remote waters. The downside of this design, however, is that the only locations you know are where you applied the tag and where it came off. The whale's path during the intervening weeks is completely unknown.

If deploying a tag on a swimming whale is difficult from an inflatable skiff, it is downright heroic to do it from a schooner, as Kahn was forced to do. While some researchers have used crossbows or compressed-air guns to fire tags into the blubber, Kahn wanted to avoid these invasive methods. In any case, researchers cannot carry weapons in Indonesia. That meant Kahn had to attach each tag to a bamboo pole and hope that he could bring the ship close enough to reach over the side and plonk it onto the whale's back. Finally, on May 3, after almost a week of thwarted attempts, Kahn had a magical day, successfully tagging both a blue whale and a sperm whale just four hours and less than 6 miles apart. The sperm whale lobtailed as Kahn leaned over the rail and just missed strik- ing him in the head with its flukes, but both scientist and instru- ments survived. "We had the same starting point for the largest baleen whale and the largest toothed whale. We had them together for this moment in time, and then they shot off in completely opposite directions." The sperm whale zipped southwest into the Indian Ocean, where the tag began transmitting 42 days later, almost 700 miles from its original location. Meanwhile, the blue whale moved northeast, and 60 days and 425 miles later, the tag began transmitting from the Banda Sea. Presumably, the whale trav- elled through the bottleneck of the Ombai Strait to get there.

Since the blue whale was only 400-odd miles from its starting point after two months, it was probably feeding, not migrating. While no one knows what zigzagging movements the animal made, a migrating whale likely would have opened up a greater straight- line distance over 60 days. The time-depth data from the tag was equivocal, however. The whale spent 97 percent of the nighttime hours (6 p.m. to 6 a.m.) at or near the surface, while during the day its time was about evenly split between swimming near the surface and making deep dives, presumably to feed. What was surprising, however, was that on 15 of the 60 days, the whale never dove deeper than 165 feet. That suggests travelling, not feeding. Once again, blue whales refuse to do what they're supposed to.

While Kahn is eager to learn more about the seasonal move- ments of the Indonesian blue whales, and to sort out whether they are the same ones seen off Australia, he is much more interested in protecting them from the array of threats they face in this enormous island country. Indonesia is creating a network of marine-protected areas that will restrict harmful human activities, and most of this conservation work is aimed at fragile coral reefs, mangroves and other coastal areas. Kahn would like to see some deep-sea areas protected as well. These are traditionally harder to manage—it is difficult to keep an eye on things that occur far offshore—but Kahn feels that the blue whale sites in Indonesia are the perfect place to try, because they are among a small number of places in the world where profoundly deep seas occur right near the coast. Indonesia is a constellation of volcanic islands, so there is no continental shelf, and in some places the water is a couple of miles deep within a stone's throw of the beach. Kahn feels that this highly unusual geography provides an opportunity for conservation groups to create plans for protecting deep-sea habitat without the logistical nightmares that usually brings.

What do the Indonesian blue whales need to be protected from? Kahn says they may be at risk from ship strikes, especially if they spend a lot of time milling near the surface at night, as his tagged whale did. The whales are often within 500 feet of the shore, which may make them vulnerable to gill nets and drift nets. The blues also face some threats that are uniquely Indonesian, although Kahn admits no one really knows the extent of the harm these may cause. The first concern is an illegal technique known as reef bombing, in which fishermen fill bottles with gasoline and fertilizer, stopping them with waterproof fuses, then lighting them and tossing them into schools of fish. "Local people don't do it much—they blame it on outsiders who roam around Indonesia bombing other people's backyards," Kahn explains. "The most remote reefs are actually the most damaged, because with no community claiming them, they get bombed to bits." The explosions are fairly small, and one would likely need to make a direct hit on a blue whale to cause any physical harm, but Kahn worries about damage to the habitat. Noise from the repeated blasts carry well beyond the area where the bottles are dropped, and may discourage blue whales from using certain migratory corridors. Kahn and his colleagues have managed to stop reef bombing around Komodo National Park, and the result has been more baleen whales in those waters. "In 2006 we had two or three blue whales coming right through the park. Nobody's ever seen that before in living memory."

The most exotic of all the potential threats to these whales—and indeed to any population of blue whales—comes from the village of Lamakera on Solor Island. Lamakera has a long history of harpooning baleen whales from traditional boats, and while this practice has been rare in the last decade, the people continue to hunt when the opportunity presents. They call their quarry *kelaru*, but it is unclear which species of whale this refers to. One candidate is the little-known pygmy Bryde's whale (Kahn recently identified one near Komodo), but since 2001 the only baleen whales seen in the vicinity of Lamakera during Kahn's surveys have been blues. Is it possible that the *kelaru* have been blue whales all along? Might this tiny Indonesian island be the only place on earth where humans hunted the largest of all sea creatures without modern technology? Could the Lamakera villagers still be taking the occasional blue whale? Kahn considers the possibility real enough to include in his annual activity reports.

It's a compelling story, one that conjures images of impossibly brave islanders accomplishing what European and American whalers could never do. One can imagine them being pulled along by a giant blue whale in a David-and-Goliath struggle, a "Lamakera sleigh ride." Traditional whalers from the similarly named village of Lamalera, on a nearby island, still hunt sperm whales from small boats, and in 2004 managed to kill a dozen of these extraordinarily strong creatures. (Indonesia is not a member of the IWC, so it is not bound to respect the restrictions on subsistence whaling.) Still, the idea is tough to swallow. It is difficult to fathom how anyone could ever have successfully hunted blue whales with traditional boats and harpoons. If it were possible, it seems likely that whalers in other parts of the world also would have succeeded. Today, the hunters of Lamakera have sold many of their traditional whaling boats to museums, and a younger generation has turned to fishing for manta rays. The truth about the *kelaru*, like so many aspects of indigenous cultures around the world, may be lost forever.

During the austral summer of 1997–98, the SOWER program turned its attention away from the Indian Ocean and on to the coast of Chile. This area had been of minor importance to whalers, though vessels did sometimes travel the length of the continent in pursuit of Antarctic whales, some of which moved from the Southern Ocean in summer to tropical waters in winter. In the early 20th century, for example, European whalers occasionally journeyed from the South Shetlands around Cape Horn to the Chilean fjords. This trip through the Drake Passage, where waves commonly reach 30 feet, could be hellish, and one expedition stands out as particularly grim. In 1912, a crew of 23 aboard a Norwegian ship ran into a terrible gale that drove the vessel onto shore. Only three of the whalers returned home, although many apparently survived the initial shipwreck only to be beaten to death and devoured by cannibals. Some of the natives' kayaks were later found to contain European boots stuffed with human flesh.

Half a dozen years earlier, the waters of southern Chile had been scouted by another Norwegian whaler who found the remote region promising, but lacked the funds to set up shop. By 1909, he had convinced an investor to sponsor an expedition, and a ship called the *Vesterlide* set out from the South Shetlands after the Antarctic season had ended. The vessel managed to make it around the cape without any of the crew being eaten, and then headed up the coast to Chiloé Island and the Gulf of Corcovado. The crew had heard reports of blue whales in the gulf, and that is exactly what they found, albeit not in the great numbers they had hoped for: between May and October that year, the whalers killed 37 blues. The Norwegians also set up a shore-based operation in Valdivia, about 150 miles north of Chiloé, but it was never very profitable and it closed in 1913.

Whalers continued to take small catches of blue whales off Chile into the late 1960s. From 1964 to 1968, in fact, a Japanese company operated in the area with consent from the government of Chile, which was not yet a member of the IWC. Hiding behind this arrangement, the whalers killed 365 blues in 1965, the biggest single-year haul ever in that region, and took 65 more in 1966–67, after the IWC's global ban. Researchers aboard these whaling ships shot Discovery-type marks into a couple of blue whales in December 1966, but never recovered them, so they learned nothing about the range of these animals. Indeed, precious little was known about this southeast Pacific population even in the late 1990s. While whalers initially believed the area was part of a winter route for Antarctic blues, the animals were also present in summer. Were they indeed Antarctic blues that ventured up the coast of South America, or were they part of a group that migrated between southern Chile and the equator? If they were not Antarctic blues, did they belong to the same pygmy subspecies as the blue whales in the Indian Ocean? Some scientists have argued that they might be some of each. One Chilean scientist reported that between 1965 and 1967 "we identified, among 168 blue whales examined, 10 specimens of the subspecies pygmy blue whale."[13] Unfortunately, the author provided no data about these specimens. What were their body measurements? How did he know some were pygmy blues and not simply juveniles or small adults? That was not clear, yet this paper continues to be cited by scientists, and even decades later the belief has persisted that both subspecies inhabit Chilean waters, an example of how dubious information can become entrenched when it is cited uncritically in the scientific literature.

One of the reasons that so little is known about the blue whales off Chile is that they concentrate in an isolated part of the country. Although Chiloé is the fifth-largest island in South America, it is more than 600 miles south of Santiago and Valparaiso, Chile's cultural centres, and it has only a small human population. It simply is not an easy place to send a team of whale biologists. The American scientists William Cummings and Paul Thompson made their historic recording of blue whale calls in this area in 1970 from a well-equipped 125-foot vessel, but for almost three decades after that, no significant research was conducted on the blue whales here. Which brings us back to the SOWER survey that started down the coast in December 1997. One of the researchers aboard that cruise was Rodrigo Hucke-Gaete, a Chilean undergraduate student whose main research subject was fur seals. After his first encounter with the world's largest animal, however, his focus soon changed. "After

I saw my first blue whale I thought, OK, now I can die," he remembered years later.[14] Within two years Hucke-Gaete had become one of the key figures in the effort to understand and protect the blue whales of his native country.

The two ships involved in the SOWER cruise spotted 47 blue whales during their initial survey. Then, just days after the cruise wrapped up, one of the researchers sighted more than 60 blues in a single day near southern Chiloé. There were a few scattered sightings after that, including nine by Hucke-Gaete personally in 2001. Two years later, with the assistance of a newly founded nonprofit group, the Centro Ballena Azul (Blue Whale Centre), he secured the funding he needed to do a proper survey of the area, and the blue whale population turned out to be much denser than anyone knew. From January to April 2003, Hucke-Gaete and his team flew five aerial surveys within 25 miles of the coast, as well as chartering a ship to run line transects. They covered the Gulf of Corcovado, the western coast of Chiloé and the waters surrounding the smaller islands to the south. In three and a half months, Hucke-Gaete noted 153 blue whales, including at least 11 cow–calf pairs. Many of the whales were feeding, and some were within half a mile of shore, a situation that occurs in only a few places on earth. When he announced his find, Hucke-Gaete called the area "the most important blue whale feeding and nursing ground discovered to date in the Southern Hemisphere."[15]

Since that groundbreaking survey, Hucke-Gaete and the Centro Ballena Azul have returned to the Gulf of Corcovado and Chiloé region every summer, working out of Melinka, a fishing village on Isla Ascensión. Just getting to their study area is a full-fledged excursion; the road ends in Puerto Montt, more than 100 miles north of Melinka, and the team has to haul its gear by barge from there. Since 2004, most of the surveys have been done from a rigid-hulled inflatable, aptly named the *Musculus*, and the whales come so close to shore that researchers rarely have to venture out more than 12 miles. They are directed by observers who scan the sea with

BLUE WHALE FEEDING AREAS IN SOUTHERN CHILE

CHILE

Isla Mocha

Valdivia

Puerto Montt

Pacific Ocean

Chiloé

Gulf of Corcovado

Melinka

Isla Ascension

Taitao Peninsula

ARGENTINA

Atlantic Ocean

N

0   100 Miles
0   100 KM

binoculars from the top of a hill; when the spotters see a blow, they radio to the boat crew, which speeds off to get a photograph. By 2007, they had identified about a hundred individuals, but had not yet seen the same animal in their study area in different years.

In addition to building a photo-ID database, Hucke-Gaete has been quick to embrace newer technologies, and in February 2004 he invited Bruce Mate to bring his satellite tags to the area. They made five successful deployments, and the tags transmitted their locations for 46 to 203 days with some interesting results. Two of the animals ventured some 1,200 miles north to the Nazca Ridge off northern Chile, which is believed to be a region of more or less permanent upwelling. Like the Costa Rica Dome, it could be both a breeding area and an area where blues can feed all winter. A third whale, a mother with a calf, headed north to the waters off Mocha, an island once frequented by British and Dutch pirates. (The white sperm whale that inspired *Moby-Dick* was seen in this region.) "She was milling around that island for a couple of days, and then she started going south," Hucke-Gaete says. "She reached an area a little south of the Taitao Peninsula, and then she rocketed north toward the Nazca Ridge."

The Centro Ballena Azul is not the only research group working in the region. Since 2004, Bárbara Galletti and her colleagues at the Centro de Conservación Cetacea (CCC) have been surveying the waters off northwestern Chiloé, about 125 miles from Hucke-Gaete's study area. During their field season from February to May, the blues come so close to shore the researchers can do much of their observing perched on a platform 300 feet above sea level. "Usually at least once a day you can see them really well from the cliff, less than one mile from shore," Galletti says. The CCC also has a small boat for their photo-ID work, and in 2006 they did an aerial survey with an aircraft on loan from the Chilean navy. Galletti's team has compiled a catalogue of more than 140 blue whales, and have had several resightings. On March 21, 2006, they spotted a blue whale that had first been photographed the previous February; the two locations were only 8 miles apart. Then in 2007 they had a banner year, with 16 resightings from previous years off Chiloé, as well as one that had earlier been seen in the Gulf of Corcovado. These findings confirmed what Galletti had suspected all along: that the same blue whales were returning to this rich feeding area in the Chilean fjords year after year, and that they were browsing between several sites, including her own study area and Hucke-Gaete's to the south. As in every other blue whale feeding ground, the oceanography in this region of South America creates just the right conditions for krill. The sea here is rather like an estuary, fed by freshwater runoff from glaciers, rivers and heavy rains. The tides and the contours of the seafloor and coastline trap the phytoplankton blooms, which spawn masses of *Euphausia valentini*, the same krill species preyed on by Ichihara's pygmy blues.

All of this led Galletti to wonder where else these whales might be travelling as they moved through the seasons. She sent her photo-ID catalogue to John Calambokidis, who compared it with his photos from the eastern tropical Pacific. This is a possible winter destination for the Chilean blues; after all, their acoustic calls have been heard on hydrophones right up to the equator. Calambokidis found no matches, but both catalogues are still relatively small, and it is possible that a match will be made in the future.

Where do these blue whales fit in the taxonomic tangle? The current scientific consensus is that every blue whale south of the equator is either an Antarctic blue or a pygmy. But the Chilean whales do not fit with the description of either subspecies. They're almost certainly not Antarctic blues: they are acoustically distinct, and there is no evidence they get much farther south than 45°S. Although the region was visited by whalers who believed they were following Antarctic whales north, it seems clear now that the whales they killed off Chiloé and in the Gulf of Corcovado were simply the ancestors of the population that lives there today. Blue whales in the Chilean fjords are usually observed from December through May, peaking in the austral summer, when Antarctic blues would likely be much farther south. For this reason, they have been labelled pygmies by default. The problem is that they are considerably bigger than the blue whales in the Indian Ocean. First there is the 1974 report claiming that 10 out of 168 blues examined in the 1960s were pygmies, which implies the other 158 were too large to be considered *B. m. brevicauda*. Indeed, while that report does not provide data for the individual whales examined, it does say that the largest one was 84 feet, far longer than any pygmy blue whale ever observed. In 2005, a blue whale stranded on the shore of Chiloé and the Centro de Conservación Cetacea measured it at 80 feet, also longer than any known pygmy blue—and it was a male, the shorter of the two sexes. Hucke-Gaete says he can get a rough estimate of a whale's length by comparing it with his 23-foot boat, and he regularly encounters animals that are three and a half times as long. So if the Chilean whales are larger than pygmies, but they are not Antarctic blues, where do they fit in? Could they be a subspecies all their own?

This question looms largest in Chile, but there are other lingering uncertainties about Antarctic and pygmy blues. Does it make sense to classify all southern hemisphere blues as pygmies if they feed outside the Antarctic in summer, or is this simply a label of convenience? Are the populations in the Indian Ocean and the southeast Pacific even closely related? To tackle these problems, researchers looked not only to acoustics, which helped answer similar questions in the northern hemisphere, but to another field that promised more definitive answers. These scientists believed that the final word on subspecies would be found in the animals' genetic blueprints. They began to peer inside the blue whales' DNA.

TEN

BLUE GENES

EVEN ACOUSTICS EXPERTS would concede that the ideal way to separate breeding populations of blue whales is with a microscope, not a hydrophone. If you hear a man with a Brooklyn accent and a woman with a Texas drawl, you can be quite sure they grew up in different places, but you cannot conclude that they are unrelated. They may have had a common ancestor 2, 3 or 20 generations back—they might even be twins who were separated at birth. Using acoustic signatures alone to delineate populations of blue whales, as useful as that has proved to be, is also limited. Call patterns can tell you where the animals spend most of their time and who they're communicating with, but they do not tell you whether there has been any mixing of populations in the past, or whether any cross-pollination is going on now. DNA, however, doesn't lie.

That was on Carole Conway's mind when she began seeking out blue whale tissue samples back in 1993. Conway grew up in New Jersey, earned a bachelor's degree in marine science from Richard Stockton College and later got her master's at San Francisco State University, where she launched her work on blue whale genetics. By the time she received her PhD in ecology from the University of California Davis in 2005, she had spent more than a decade gathering and analyzing blue whale DNA from around the world. "It's fundamental to blue whale conservation to know their breeding populations and their stock structure," Conway says. "I used to hear all the time that there are 10,000 blue whales left in the world. That figure is meaningless unless you know the population structure." Her point is that you cannot get accurate population estimates—and therefore cannot tell whether blue whales are recovering—unless you know the geographic boundaries of their breeding stocks. "How can you take a census of Los Angeles, for example, if you don't know where the city limits are? When you collect demographic data for blue whales, you need to be clear about both geographic and temporal bounds; because these whales migrate, you have to know where they are at what times of year." Scientists have long recognized the importance of drawing clearer boundaries through genetic analysis, but before Conway no one had done a comparative study with global reach.

The eastern North Pacific blues, numbering at least 2,000, appear to be thriving. The blues off Iceland seem to be increasing, too, and even the decimated Antarctic population is growing. But population estimates in other regions are typically well under a thousand individuals. The question, then, is, how many blue whales constitute a healthy breeding population? "There's no set number," Conway explains. "You worry about it when it gets really small, say, under 50 individuals, because then you lose genetic variance." That means inbreeding, which can decrease fertility and cause other health problems. "You're also concerned about running into demographic problems—maybe the sex ratio isn't equal any more, or maybe the animals can't find mates because there are so few. But it depends on the species and its life habits." For blue whales the obstacles appear high: they probably do not reach sexual maturity until about age 10, females may bear only one calf every two or three years, and they roam large expanses of ocean alone. The good news, however, is that severely depleted populations of other whale species seem to recover if they are given full protection.

Conway's first challenge was simply convincing other researchers to share their blue whale biopsies. "As you can imagine, these are precious samples. I had to earn their trust over several years." John Calambokidis and Richard Sears were both helpful, and Conway received a big boost from the Southwest Fisheries Science Center in La Jolla, California, which has the largest collection of skin and blubber samples from blue whales around the world. Eventually, Conway cobbled together 204 samples from 14 locations around the globe, albeit with uneven distribution. The best variety came from the blues off California (41 samples) and the St. Lawrence (38), with smaller numbers from Mexico, the Costa Rica Dome, Chile, the Galápagos Islands, Australia, Madagascar and the Antarctic. The only significant regions from which she was unable to obtain samples were the northwestern Pacific and the eastern Atlantic, where blues have not been studied extensively. (She did receive two biopsies from the northern Indian Ocean, but excluded them because such a small number was not likely to be representative of that population.)

With all of these samples in hand, it was time for some detective work. "The work we do is really the same as forensic work you hear about in court cases," Conway says. "It targets a particular part of the DNA—in my case, there were five specific markers—and we look for what are called allele frequency differences." An allele is one of a number of possible states that a given gene can have; a gene that determines eye colour would have an allele for blue, another for brown, another for green, and so on. "Genetic analysis looks at how often these alleles occur in a population. In Norway, for instance, you're going to find the allele for blue eyes much more often than the allele for brown, whereas in China you're going to find the opposite. When the frequencies are significantly different, that indicates they are different breeding populations." Conway grouped her blue whale samples by geographic area and zeroed in on her five markers: not the eye-colour gene, but other regions of the DNA that would reveal common heredity. "If there are big enough differences between the samples from, say, California and Chile, then you can tell those whales are not breeding together."

Conway stresses that 204 samples cannot be the final word on global population structure, but her years of work have painted the map with some broad brushstrokes. She was able to identify four subsets that roughly correspond to different ocean basins: the north-east and eastern tropical Pacific, the southern Indian Ocean, the western North Atlantic and the Southern Ocean. That means there are at least four breeding stocks of blue whales, and that little or no commingling goes on between members of these four groups. However—you knew this was coming—nothing about blue whales is that simple. And, again, the wrench in the works comes from that enigmatic group of blues off Chile. Conway was unable to determine whether these whales constitute a fifth population. The Chilean whales were genetically distinct from those in the Antarctic, but curiously, they were not significantly different from those off California and Mexico, nor from those in the Indian Ocean. Part of the reason these groups looked similar may have been the small sample size. Had Conway been able to look at hundreds of biopsies from each area, differences may have emerged. But that doesn't mean the similarities are insignificant. Like most researchers, Conway believes that there are indeed two distinct populations off the west coast of the Americas—a belief backed up by acoustic data—but her analysis showed they may be closely related. In fact, blues in northern and southern regions of the eastern Pacific may interbreed occasionally, not only at the equator, but also in the Costa Rica Dome.

Bruce Mate's satellite tags, Kate Stafford's hydrophones and John Calambokidis's cameras all found clear links between the California population and the Dome in the 1990s. But while several researchers had suggested the Dome might be visited by southern hemisphere blue whales as well, no one had been able to confirm it. Stafford's acoustic data suggested that blues from both hemispheres probably do overlap at the equator, but she found little evidence of this in the Dome, which is more than 600 miles farther north. Conway's data, however, suggested that mixing may occur in both places. One allele, for example, was common in whales from California to the Galápagos, but absent in Chile, while another was moderately common from Chile to Costa Rica, and occasionally present as far north as Mexico, but not in California. The fact that these two alleles overlap at the equator and in the Dome may indicate some exchange. "I can't say for sure whether they're breeding together or they're simply there at the same time," Conway says, "but it looks like the whales off Costa Rica as well as the Galápagos show some characteristics of both hemispheres." Conway also found more genetic variation among the whales in the eastern tropical Pacific than among those off California, even though the former sample was much smaller. "When you see big differences within one group, it might indicate that it's actually two groups that have come together."

Whether blue whales from different hemispheres interbreed isn't just a question of geography, but also one of timing. If the animals mate exclusively during their respective winters, then the two populations would be six months out of phase: southern whales would be looking to mate from about June to August, while those from the north would be on the prowl from about December to February. Even if they did happen to meet at the same time, there might not be mutual interest in romance. "When I started out," Conway says, "I remember other researchers telling me, 'Oh no, because the whales from each hemisphere are out of sync, interbreeding never happens.' Well, I think it does happen sometimes, and more than we thought." If their migrations show only trends and not strict timetables, and they feed year-round, perhaps blues occasionally mate in the off-season, too. Indeed, because the Costa Rica Dome and the waters west of the Galápagos are both upwelling areas with abundant krill, which is unusual in tropical waters, they are likely places for mixing to occur. "This may not be happening all over," says Conway. "It may be unique to the eastern tropical Pacific. Because northern hemisphere blue whales are able to feed on their wintering grounds as well as their summering grounds, maybe there's not so tight a deadline for them to migrate. Maybe some of the whales from the north can afford to go to the Costa Rica Dome and the Galápagos early because they can feed there, too. And maybe some of the whales are going to migrate down toward Chile, but they don't have to leave as soon. The samples I had from the eastern tropical Pacific, interestingly enough, were from November, and that is a time of transition for both their migrations. What we need to do is get samples from other times of year as well, and see if those samples are dominated by whales from one hemisphere or the other. Then we can learn whether whales from the two hemispheres really are using these areas, and what months they overlap."

There was another tantalizing piece of genetic evidence to support the idea of interbreeding. Two whales from the eastern Pacific had DNA containing an allele common in the northern hemisphere and another that was common in the south—all the others contained one or neither, but never both. One of these individuals was sampled off the Galápagos, the other in the Dome. In some ways, this finding recalls the blue whale in the Gulf of Alaska whose calls had characteristics of both eastern and western populations, but with one important difference. While that bilingual animal might have been the offspring of parents from different populations, it might also be an eastern animal that learned how to make western calls. Sharing DNA, of course, doesn't work like that. The only way alleles can pass from one population to another is through interbreeding.

Things got even more interesting when Conway looked at her samples from the southern Indian Ocean, which were all believed to be pygmy blues. This included six whales sampled near Madagascar as well as 23 from southern and western Australia. Although these whales were on opposite sides of an ocean basin, their alleles were very similar. At the same time, they were clearly distinct from the Southern Ocean samples, meaning that these whales were not interbreeding with Antarctic blues. "That surprised me, because there's no obvious geographical feature separating them," says Conway, who expected at least some gene flow between these groups. After all, she found that to be the case with the overlapping populations in the Pacific. "I think the blue whales off Antarctica are following the food along the ice edge, and because that's very productive they're not likely to take a chance and go farther north looking for other feeding grounds. And I think they choose their mates from who they feed with. That might be the reason why there's not so much of a difference in the eastern Pacific, but a big difference between the Indian and the Southern Oceans."

The real oddball in Conway's collection was a whale biopsied off Chile in the late 1990s. This animal's DNA contained an allele that was not found in any of the 15 other samples from Chile, but was common in the Indian Ocean. "That was like a smoking gun—something's going on there," Conway says. She's quick to point out that the whales off Chile are almost certainly distinct from those off Australia or Madagascar, but there must have been some connection in the past. Perhaps there was what geneticists call a "founder event," in which a few individuals leave one group and start a separate population in a different area. Maybe one or more whales from the Indian Ocean wandered into Chilean waters and stayed there; this might still happen now and again. "With genetics we have a hard time pinpointing when events like that happened." Genetic signals are a bit like digital computer files: they can be copied through many generations without deteriorating. So while allele frequency differences can tell you whether two populations show signs of mixing, they cannot determine whether it happened recently or in the distant past.

Conway's research was some of the first to arrange the blue whale puzzle pieces in the southern hemisphere, though it raised as many questions as it answered, as all good science does. "This study at least shows some of the range of genetic diversity of blue whales. If you want to conserve the species, and as much variation as possible within the species, then you'd want to make sure you conserve some in each of these areas. Whether these populations are different subspecies is another question that I'll leave to someone else."

Using genetics to solve the pygmy–versus–Antarctic blue whale question was already being tried by someone else. An important aim of the SOWER research cruises was, after all, to determine a reliable and nonlethal way of distinguishing between these two subspecies. Before Carole Conway's study was completed in 2005, it had long been suspected that although *B. m. brevicauda* and *B. m. intermedia* have distinct summer ranges—the boundary separating them is around 55°S latitude—some individuals of each subspecies occasionally roam into the territory of the other. Exactly how many or how often, however, wasn't known, and until the question was resolved it would be next to impossible to assess the health of either population.

Early in the last century, the oceans held about 240,000 Antarctic blue whales, while the number of pygmy blues throughout the hemisphere was a small fraction of that. However, Antarctic blues were far more heavily depleted than their smaller relatives. As a result, pygmies probably now outnumber Antarctic blues significantly, and if the subspecies still overlap on the Antarctic feeding grounds, the proportion of pygmy blues could be much higher than it was during the whaling era. That would make accurately counting Antarctic blue whales and assessing their recovery impossible— the overall population in the Southern Ocean might appear to be increasing, but most or all of the growth might be pygmy blues. The upshot was that the IWC needed to know what percentage of blue whales sighted in the Antarctic were pygmies. Even as late as 2003, most everyone agreed that it was less than 7 percent, but that wasn't precise enough.

There was a good reason for the confusion. Tadayoshi Ichihara's original description of pygmy blues was based on the careful measurement of more than 300 carcasses. It is relatively straightforward to distinguish between pygmies and Antarctic blues if you kill and examine them, but this was obviously no longer an option, and it wasn't clear whether anyone could actually tell the difference from the deck of a ship. Certainly there were people who claimed they could—the SOWER cruises included experienced Japanese crewmen whose job was exactly that—but no one knew how reliable these observers were. It is extremely hard to visually estimate a whale's length with any precision, and length alone often tells you nothing, since the two subspecies overlap in size. Ichihara described a number of morphological differences other than length, but these too were far more easily detected in dead whales.

So in the mid-1990s the IWC charged its Scientific Committee with a new task. "They wanted to know whether we could tell them apart genetically," says Rick LeDuc, a geneticist at the Southwest Fisheries Science Center in La Jolla, California, who was part of the original team. "If you take a biopsy, they asked, can you tell whether it's a pygmy or a true blue? We said we'd give it a shot." LeDuc and his colleagues got tissue samples from around the Antarctic continent and from several locations in the Indian Ocean: off western Australia, south of Madagascar, even from the Maldives, north of the equator. "And just to be thorough, we included samples taken from the southeast Pacific, too—off Chile, Ecuador and Peru— because it was thought that some of these blue whales occasionally ventured down to Antarctica, too."

Unlike Conway's study, which looked at overall genetic trends, LeDuc's goal was to find a diagnostic marker—a specific region of DNA that is always the same in any given group, and always different from other groups. But when the geneticists examined their 111 biopsies, they found nothing that consistently distinguished pygmies from Antarctic blues. It would have been easier if they had what geneticists call holotypes. These are specimens, often from a museum, that can be used as benchmarks, as in: "Here is a tissue sample that we're certain is from an Antarctic blue, and here's another we know for sure is a pygmy. Now compare the rest to these." But no one has a genetic sample from a southern hemisphere blue whale that has also been comprehensively examined and measured to confirm its subspecies. That, of course, would involve killing the animal. All LeDuc's team knew was where each animal was biopsied. "And we didn't know how important the geographic origin of the sample was. If there are a lot of animals wandering from one area to another, then the geographic area doesn't mean anything. If we find a genetic marker once in the Antarctic and once in the Pacific, does that mean that it occurs naturally in both populations, or was this an animal that moved from the Antarctic to the Pacific? That was the conundrum we faced: how do we interpret the differences we found?"

LeDuc's compromise was to cast a wider net and look at frequency differences in a number of genetic markers, similar to what Conway did in her global study. This method cannot tell you with certainty whether a whale belongs to one group or another, but geneticists can use it to assign probabilities: you may be able to say that sample A is 80 percent likely to be an Antarctic blue whale and 20 percent likely to be a pygmy. But even that raised problems, LeDuc says. "Coming up with probabilities was more difficult than we thought because the two groups we were lumping together as pygmies—the Indian Ocean and the southeast Pacific—are not even that similar to each other." The whales off western South America, in other words, were just as different from the pygmy blues in the Indian Ocean as they were from Antarctic blues—there were three broad groupings now, not two. The analysis also turned up two whales that were potential vagrants. One of the blue whales was biopsied in the Antarctic, but the genetic analysis indicated that it was 94 percent likely to have come from the southeast Pacific. A second animal sampled in Chilean waters was 92 percent likely to have come from the Antarctic. "And for what it's worth," LeDuc says, "there were animals from the Antarctic that were identified as pygmies from the ships that are unlikely to have been pygmies based on our genetic analysis. So either the genetic analysis didn't pick up something, or the guys on the ships were wrong."

The guys on the ships were now back at square one, as LeDuc and his colleagues eventually had to admit that a biopsy was not going to determine the subspecies conclusively. The search was on again for some physical difference—and it had to be something in plain view—that could be used to tell the subspecies apart. During many of the SOWER cruises, the crews recorded video of blue whales, and a team led by the Japanese scientist Hidehiro Kato set about examining these for clues. By carefully studying more than 28 hours of video, Kato confirmed that those found north of 55°S in summer, and therefore presumed to be pygmy blues, have a proportionally larger head and shorter tail: he called it a "tadpole shape" as opposed to the Antarctic blue's "torpedo shape." He also made a couple of interesting new observations. For one, he noticed that the animals' blowholes are different: in the pygmy blue, the groove in the centre usually extends past the whale's nostrils, while in the Antarctic blue it does not. Finally, pygmy blues almost always have a hump or series of ridges along the back, while Antarctic blues almost never do. Though none of these was a slam dunk on its own, if observers could note all three characteristics on a given individual, they could combine them to come up with an accurate probability, Kato argued. If a researcher photographed a whale deemed to be torpedo shaped, with a blowhole whose central groove did not extend past the nostrils, and with no visible dorsal hump, the chances that it was an Antarctic blue would be 99.8 percent. Even with only two of the three characteristics confirmed, the probabilities were often greater than 92 percent one way or the other. It wasn't a perfect method, but it was promising—as long you could get clear photos of the whale you wanted to identify.

Other researchers looked to acoustics. Even in the mid-1990s, before anyone had confirmed that blue whale populations use distinctive calls, the SOWER crews routinely dropped sonobuoys all over the southern hemisphere. These recordings, combined with those by other researchers in the decade that followed, have once again shown that acoustics may be the single most useful tool for understanding blue whale demographics. The 1996–97 SOWER program consisted of two cruises, the first of which travelled to an area south of Madagascar, where Japanese vessels had earlier sighted what they believed were pygmy blues. The second leg took the researchers right to the ice edge in the Antarctic. There were significant differences between the two groups of blue whales they encountered. First, the shipboard observers identified virtually all of those off Madagascar as pygmies, while all of those in the Antarctic study area were deemed to be true blues. The two groups also made unique vocalizations. The Madagascar whales made a two-part call: a pulse around 38 Hz, followed by a pause of about 25 seconds, and then a long downsweep. The animals in the Antarctic, however, repeated a single call over and over—it began at 28 Hz, then swept sharply down to about 20 Hz. This gave scientists their first inkling that acoustics, surprisingly, might end up succeeding where genetics had fallen short.

Later SOWER cruises to the Antarctic from 2001 and 2003 dropped acoustic recorders that remained in the sea year-round. All of the blue whales picked up by these hydrophones emitted that same single-unit call beginning at 28 Hz. These calls have since been heard in every month (peaking in March and April, and again in October and November), all around the Antarctic continent. Meanwhile, none of the sounds attributed to pygmy blues off Australia or Madagascar have ever been heard south of 60°S. The acoustic evidence, then, strongly suggests that pygmy blues virtually never move into the Antarctic after all. When the SOWER program began, about 1 in 50 blue whales—and perhaps as many as 1 in 15—in the Southern Ocean were estimated to be pygmies. It was starting to look like that number was a lot closer to zero.

Even after it became clear that Antarctic blue whales make consistent calls, the acoustic picture was still fuzzy in the Indian Ocean and the southeast Pacific. It was Kate Stafford who eventually tackled this problem. She compared the recordings made off Madagascar with more recent ones made off southwestern Australia and found them to be distinct—the Aussie whales made a three-part call, as opposed to the two-part pattern heard on the opposite side of the ocean. Then she dug into the archives and found the earliest recordings of blue whales off Sri Lanka, which were made in the early 1980s, when the only people who knew anything about blue whale sounds were in the U.S. Navy. These blues made a four-part call. That meant there were at least three separate breeding populations of pygmy blue whales in the Indian Ocean.

As for the whales found in the southeast Pacific—the ones that LeDuc's team and many other scientists also assumed were pygmy blues—they too made distinctive calls. Recordings in this region, first made by Cummings and Thompson off Chile in 1970 and picked up again on bottom-mounted hydrophones off Peru in the 1990s, were different from those of any of the Indian Ocean whales, as well as from the 28 Hz calls heard in the Southern Ocean. That was pretty good evidence that the blues off Chile had not migrated up from the Antarctic. But if they were pygmy blues, then they must be a *fourth* breeding population, separate from the three noted by Stafford in the Indian Ocean.

LeDuc's research was bringing him to the same conclusion. His continuing genetic analysis convinced him that the blue whales in the southeast Pacific were not closely related to any others in the southern hemisphere. "What we found, which agrees with the acoustics, is that the southeast Pacific blue whales are no more similar genetically to the Indian Ocean pygmies than they are to the Antarctic blue whales. They're all about equally different from the others. That brings up the question of whether the ones in the southeast Pacific should be considered pygmy blue whales. I think the new issue is, are those whales off Chile a separate subspecies all their own?"

Like the poem about the blind men and the elephant—each one touches a different part of the animal's body and arrives at a wholly different conclusion about what it is—a scientist studying one aspect of blue whale biology or behaviour may have a vastly different perspective than one in another. Often what is needed is someone who can weave a tapestry from the threads of everyone else's work. For the blue whales of the southern hemisphere, that someone has turned out to be Trevor Branch. Born in South Africa in 1974, Branch collected bachelor's degrees in zoology and computer science and a master's in conservation biology in Cape Town before moving to the U.S. and earning a PhD at the University of Washington.

He is quick to admit that he doesn't do much fieldwork—he jokes that he's a "desk biologist" and has never even seen a live blue whale. But he has spent years tracking down century-old journal articles and whaling records, challenging unsubstantiated claims, and trying to synthesize the current research of field biologists, acoustics experts and geneticists. In 2007, he published an exhaustive paper that lists 41 co-authors from 17 countries—including Kate Stafford, Peter Gill, Margie Morrice, Benjamin Kahn, Rodrigo Hucke-Gaete and Bárbara Galletti—and cites an eye-glazing 220 sources. His goal was to map both the past and present distribution of southern hemisphere blue whales, and he's come as close as anyone to finally doing so.

Branch has also contributed original research that has shed new light on the question about what proportion of blue whales in the Antarctic are pygmies. His insight was to concentrate on data gathered during the whaling era rather than the present day. The advantage of this approach comes from sheer volume: whereas Carole Conway examined 204 tissue samples and Rick LeDuc looked at 111, Branch had data on more than 300,000 blue whales. Since 1980, the IWC has been gradually compiling decades of whaling statistics in a massive database that includes upwards of two million animals. This mountain of information includes the length, sex, pregnancy status and location of almost every blue whale killed between 1913 and 1973. Add that to the thousands of sightings, strandings, Discovery marks and acoustic recordings, and Branch was awash in data. He asked himself how he could apply all of this to the problems other scientists were wrestling with: What are the historical and current ranges of pygmy and Antarctic blue whales? How many breeding populations are there in the southern hemisphere? Are the subspecies valid, or should they be reconsidered?

The first issue Branch tackled was what proportion of blue whales killed in the Antarctic were pygmies. Although it would have been relatively straightforward to determine the subspecies of a dead whale, pygmy blues were not identified until the early

1960s, long after the golden age of blue whale hunting. Ichihara had argued that the so-called *Myrbjønner*—the smaller-than-average blues occasionally taken at South Georgia and elsewhere—were pygmy blues, but that had never been verified. Decades after the fact, with no carcasses to measure, how could it be? Branch decided to zero in on the whaling records for sexually mature females, since that's where size overlap between the subspecies is least likely. The average length of a female Antarctic blue at sexual maturity is almost 78 feet, and only a tiny percentage of pygmies ever gets that big. In a typical population, then, very few Antarctic blues would reach sexual maturity before 74 feet or so, while immature pygmy blues over that length would be equally rare. If you knew a female blue whale's length and whether or not it was mature, you would have a good idea which subspecies it belonged to. But unless the whale happened to be pregnant when it was killed, how would whalers have known whether it was sexually mature?

Fortunately, early biologists had discovered a way. When mammals ovulate—in blue whales this happens about once every 30 months—the eggs leave behind knobs of tissue that eventually shrink and harden into spheres called corpora albicantia. Sexually immature females, then, have no ovarian corpora, while mature females add a new one about once every two and a half years. Branch went through whaling records and found reliable data for more than 100 females that were confirmed to be pygmy blues (they were caught in 1960 or later) and more than 2,000 that were killed in the Antarctic but had not been identified by subspecies. When he plotted each on a graph according to length and number of ovarian corpora, the data sorted itself into two distinct groups. Among the known pygmies, 70 females were between 68 and 73 feet and all were sexually mature; in the Antarctic, only 1 out of 75 whales was sexually mature at that size. Meanwhile, all of the pygmy blues between 74 and 77 feet had at least four ovarian corpora, with an average of nine; whales in that range from the Antarctic had no more than three, and three-quarters of them had none at all. Based on this analysis, Branch wrote that the number of blue whales in Antarctic waters that belonged to the pygmy subspecies during the whaling era "is statistically indistinguishable from zero."[1]

Branch later did a parallel analysis using the IWC's catch database, which includes 33,000 pregnant blues and more than 300 that were lactating when they were killed. (Although whalers were not supposed to kill nursing females, it is virtually impossible to identify them at sea; mothers with calves were also off limits, but apparently some nursing cows were encountered without their calf nearby, or the regulation was simply ignored.) With this enormous amount of data, Branch could be even more confident that his statistical model was robust, and it turned out to mesh nicely with his earlier numbers. This time he was able to show that almost all of the whales killed south of 52°S were Antarctic blues, while in the Indian Ocean virtually all were pygmies. The only thing that cluttered up the otherwise tidy findings, however, were the whales killed off Chile: the sexually mature females in this area fell somewhere between Antarctic and pygmy blues in length. Weighing this and the evidence gathered by other researchers, Branch concluded that "it seems clear that Chilean blue whales should be classified a separate subspecies."[2]

The genetic, acoustic and historical research is in agreement when it comes to the question of what proportion of blue whales in the Antarctic are pygmies: the answer seems to be almost zero. If blue whales are increasing in the Southern Ocean—and it appears they are—then this trend must reflect a genuine growth in the population, not simply a statistical error caused by inadvertently counting pygmy blues. Although it was far and away the most heavily hunted subspecies, *B. m. intermedia* may truly be on the slow road to recovery.

As for where these Antarctic blues go in winter, that's still not clear, and it probably never will be. The whales certainly do not all do the same thing—bottom-mounted hydrophones near the Antarctic Peninsula have picked up blue whale calls in every month, so not all individuals migrate every year. Some may head up to the eastern tropical Pacific in the fall and winter. It is possible that shore-based whalers in Chile killed some Antarctic blues near the coast, but it is more likely that any breeding area in the southeastern Pacific is much farther offshore, and hasn't yet been located. Other Antarctic animals may move into the central Indian Ocean or the waters off Australia or New Zealand; their distinctive 28 Hz calls have been detected in winter in each of these areas. In years past, some Antarctic blues wintered in the waters off southwestern Africa (and, to a lesser extent, southeastern Africa), where they were killed by shore-based whalers, though sightings in both of these regions have been extremely rare in recent decades.

As for the pygmy blue whale, which has courted debate and controversy for almost half a century, the subspecies likely includes at least three related populations. The first is in the eastern Indian Ocean. Although no one has yet made a photo-ID match, an acoustic link, or gathered anything definitive from a satellite tag, Branch believes that the whales that summer off southern Australia belong to the same population as those found off Indonesia. Benjamin Kahn has reported blues in all seasons, so not every individual follows the same pattern, but one scenario seems most likely: many of the whales probably move toward the equator in winter as the Bonney Upwelling dies down and krill become more abundant farther north. (Some also dip down to the subtropical front when the eating is good.) The second group, which is acoustically distinct from the Australian whales, moves between the waters south of Madagascar and the Subantarctic. It is not known whether these two southern Indian Ocean groups ever interbreed; their different calls suggest they don't, not to mention that their feeding sites are separated by thousands of miles. Yet Carole Conway's genetic work—with an admittedly small number of samples—did not find significant differences in their allele frequencies. Even if the blues are no longer commingling, they must have done so in the past.

What of the northern Indian Ocean population, including the blues seen off eastern Sri Lanka and the Maldives, sometimes called *B. m. indica*? These whales are unique in that they appear not to make significant migrations. Instead, they seem to move between locations straddling the equator. (Sri Lanka and the Maldives are in the low latitudes of the northern hemisphere, while the whales' calls have also been heard off Diego Garcia, an atoll located at about 7°S.) Branch is confident, however, that the *B. m. indica* subspecies is invalid. He believes these northern Indian Ocean whales are simply a third population of pygmy blues. He came to this opinion after carefully reviewing the data from illegal Soviet whaling in the decade or so before 1973. "The Russians went all over the northern Indian Ocean and caught around 1,300 blue whales there, and the lengths were identical to those of the pygmy blue whales in the southern Indian Ocean. The ovarian examinations were also identical, and the length[s] at maturity were pretty much the same. So in my mind *indica* is identical to the pygmy blue whale."

But there may be a new blue whale subspecies to take its place. Branch believes that the blue whales off Chile make a strong case for being a separate subspecies. Not only are they geographically isolated, genetically distinct and acoustically unique, but they are morphologically different from the pygmy blues of the Indian Ocean. While no one has yet examined all of the historical data, Branch's research revealed that 12 percent of the female blues killed off Chile exceeded 79.5 feet, while not one of the 12,000 pygmy blues killed in the Indian Ocean was that long. That is well in line with more recent observations of these whales, whose maximum length appears to be midway between Antarctic and pygmy blues. Until he has an actual specimen, Branch cannot formally propose a new subspecies name, but he's keeping one in his back pocket: *Balaenoptera musculus chilensis*. If it ends up being adopted, he will lay claim to naming the second-largest subspecies on earth.

ELEVEN

WHY SIZE MATTERS

JUST ABOUT every book and article that mentions blue whales repeats the claim that the animal can reach 100 feet. It's a nice round number, and the third digit imparts an extra dash of dramatic effect. The truth is, though, that the figure is highly misleading, like a reference book saying that humans may exceed 8 feet. Only about one in every 2,500 blue whales killed in the whaling era was reported to be 100 feet, and even then, some of those claims were probably fish stories.

For the record, no blue whale in the northern hemisphere has ever been known to approach that length. The longest in the North Atlantic was killed in the Davis Strait and measured 92 feet, while the biggest in the North Pacific whaling records was a mere 88 feet. Most of the blues encountered by Richard Sears and John Calambokidis today are in the low 70s, with the largest females approaching 80 feet. The pygmy blue whales of the southern hemisphere are similarly sized—generally about 70 feet, with none measured at more than 79. There are, however, many whaling-era reports of Antarctic animals that exceeded the century mark. The *Encyclopedia of Marine Mammals* lists a 104-foot specimen taken at South Georgia, and a 107-footer killed off the South Shetlands. George Small's *The Blue Whale* also reports a 107-foot animal from South Georgia in the 1920s. The largest blue whale in the IWC's database was killed on February 16, 1919, and was recorded at 115 feet. Whether these measurements were scrupulously made is impossible to know—whalers had a vested interest in exaggerating the size of their catches, either because they were paid incentives or simply because they were motivated by the prestige of landing the largest trophy, something any sport fisherman can relate to. Even if all the attempts were honest, inconsistent methodology would have produced different numbers. The accepted standard is to start at the rostrum, or snout, and measure in a straight line to the notch of the tail. (This convention came about because the flukes were typically cut off.) If the measurement was begun at the protruding lower jaw, the size could be exaggerated by more than 3 percent. If the tail flukes were not cut off, these were sometimes included as well. Measuring along the animal's back rather than in a straight line would also add a few feet. Indeed, any of these methods could have turned a 96- or 97-foot blue whale into a 100-footer. Many scientists are cautious about the lengths stated in the whaling records, but the claims are so awe-inspiring that it is tempting to accept them uncritically, and many people seem to have done just that. The *Handbook of Marine Mammals*, published in 1985, mentions a 110-foot female killed at South Georgia, noting that it was measured in the "scientifically correct manner," though the original source reveals only that it was taken some time between 1904 and 1920, which doesn't inspire confidence in the recordkeeping. In his book *Among Whales*, Roger Payne describes seeing a photograph taken in 1928 of "the largest animal for which we have any evidence in the universe," though he does not say what its length was, nor does he identify his source. Payne's date means it can't be the same whale mentioned in the *Handbook of Marine Mammals*. The inescapable fact is that no one knows for certain the length of the biggest blue whale.

In the 1980s, Dale Rice, one of the pioneers of blue whale research, did a rigorous examination of the whaling data to see how many of these claims could be verified. "After combing the literature and questioning whale biologists from all over the world," he concluded, "I find that the longest blue whale (measured in the standard zoological manner, from the tip of the snout to the notch between the tail flukes) that can be authenticated was a ninety-eight-foot female examined by Dr. Masaharu Nishiwaki of the Japanese Whale Research Institute in the Antarctic in the 1946–1947 season."[1] Of course, since it would have been a remarkable coincidence if Dr. Nishiwaki had stumbled upon the largest blue whale in history, it is almost certain that some blues did exceed 100 feet. Whether any really did reach 110 feet—let alone 115—we'll never know.

The maximum weight of a blue whale is even more difficult to ascertain, though it is even more singular than the animal's length. Some long-necked, long-tailed dinosaur species are believed to have stretched 115 or even 130 feet, which exceeds even the most exaggerated length of any blue whale. But nothing has ever come close to *Balaenoptera musculus* when it comes to sheer heft. The heaviest-known dinosaur is estimated at about 80 tons—possibly as much as 100 tons. By comparison, Dale Rice's research found an authenticated record of an 89-foot female killed in 1948 that had been cut up and weighed at 285,600 pounds, or more than 142 tons. "Allowing for 12 percent loss of blood and body fluids," Rice wrote, "she must have weighed about 320,000 pounds when alive."[2] Antarctic whalers reported several other blue whales in this range, including a female killed off South Georgia that supposedly weighed 190 tons. Whatever the veracity of these claims, the mass of an animal typically varies with the cube of its length; in other words, doubling the length increases the mass by eight times. Tacking another 11 feet onto the 89-foot whale Rice reported would, therefore, increase its mass to well over 200 tons.

Most people can visualize what 100 feet looks like, but appreciating the prodigious mass of a blue whale is like trying to envision a light-year. Consider this: the National Hockey League and baseball's American and National Leagues have a total of 60 teams, with an average roster of 24, for a total of 1,440 players. The average player weighs about 205 pounds. That means that every big-league hockey and baseball player huddled together on a scale would weigh about 148 tons—well within the range of a single large blue whale.

Why did blue whales evolve such immense proportions? What possible advantage does it give an animal to be heavier than three professional sports leagues?

The evolutionary history of blue whales has only recently become clearer, and there are still many unknowns. Indeed, explaining the emergence of cetaceans has been a challenge since scientists first began to study natural selection. They are among the very few mammals to have left life on land and taken up permanent residence in the ocean, and the fossil record suggests they did so rather suddenly, at least in evolutionary terms. Why would an air-breathing mammal take to the ocean full-time? And in the case of blue whales and the other rorquals, why did they adopt a feeding strategy employed by no other animal in the ocean: namely, engulfing huge amounts of seawater and prey, then expelling the water while trapping the food in fringed plates of keratin within their mouths?

The earliest-known archaeocete, as the ancestors of modern whales are called, was a wolf-sized mammal that lived more than 50 million years ago. The first fossil fragments of this creature, called *Pakicetus*, were discovered in the late 1970s in Pakistan. When *Pakicetus* lived, the area between what is now Eurasia and North Africa was covered by a body of water known as the Tethys Sea, and it is here, a few million years after the last dinosaurs, that a lineage of hoofed mammals probably began feeding on fish near a river. Eventually, these evolved into a species that entered estuaries, lagoons and finally the sea. *Pakicetus* looked nothing like a modern whale. Paleontologists were able to link it to cetaceans only because it had certain characteristics in its teeth and ears that are seen in no other mammal, but are present in all later whale ancestors.

About 40 million years ago, archaeocetes became fully aquatic and evolved the largely hairless bodies, elongated shape and, in some cases, the great size we associate with whales today. This means that the extraordinary leap from landlubber to denizen of the deep took less than 15 million years—an evolutionary eye blink. Where was the "missing link," the amphibious animal that must have been part of that evolution? This argument used to delight creationists until 1994, when scientists dug up the remains of a four-legged creature that resembled a furry crocodile with wide feet and a tail adapted for swimming. The fossil, again found in Pakistan, was about 48 million years old and was named *Ambulocetus natans*, "the walking and swimming whale." It was the first evidence that whales had indeed undergone the transformation from land-dwelling to aquatic mammal by natural selection, and that they had done so in an extraordinarily short period of time.

What about those unique adaptations of blue whales and their kin—their enormous body size and their engulfment-feeding strategy? When and why did these evolve? Around 35 million years ago, the line of archaeocetes branched into the two suborders that exist today: toothed whales and baleen whales. This split probably came about after the breakup of the supercontinent in the southern hemisphere. The land masses that are now Antarctica and South America broke away and opened the expanse of sea we now call the Southern Ocean. Prevailing winds began to swirl this water west to east around the new continent at the bottom of the world, and the Antarctic Circumpolar Current was born. This current, which today runs along the nether regions of the Atlantic, Pacific and Indian Oceans, probably resulted in colder temperatures, setting in motion a series of other changes. Cooler water encourages blooms of phytoplankton, which encourages krill, often in enormous numbers. This opened an evolutionary door for animals who could take advantage of this bountiful new prey source.

But while some krill patches are of mind-boggling dimensions, they are not evenly distributed over space and time. So to truly exploit prey that is occasionally superabundant but often short-lived, an animal would require some specialized abilities. First, it would need to be capable of swimming great distances in short periods, since rich feeding areas might be hundreds, even thousands, of miles apart. It would also need the ability to store energy for weeks, perhaps months, and live off its fat stores when prey was hard to come by. Finally, such an animal would require an anatomy that enabled it to ingest huge masses of tiny prey when it was available, since feeding on krill a few at a time would be far too inefficient. The common denominator in all three of these adaptations is large body size. And so it was that the hand of natural selection shaped an animal that was uniquely suited to this niche in the southern hemisphere: the first large-bodied, big-gulping, filter-feeding baleen whale. In the millions of years since, the sculptor honed this design until one species arose that dwarfed any that preceded it.

———————

It is one of evolution's remarkable incongruities that the blue whale owes its great size to an animal that is a couple of inches long. The name *krill* comes from the Norwegian word for "tiny fish" and refers to more than 80 species of what scientists call euphausiids. These are a family of small shrimp-like crustaceans that feed directly on phytoplankton, and occasionally zooplankton (tiny animals), and are themselves preyed on by numerous species of fish, mammals and birds, making them the keystone in many marine food chains. One of the most simple and elegant of these food chains is the one that goes phytoplankton to krill to blue whale. As the biologist and author Douglas Chadwick puts it, "It means that blues are only a couple of steps removed from dining directly on sunlight."[3]

When blue whales started to appear off California in the 1980s, it was quickly obvious that they were drawn to the area by the bounty of krill that was available there in summer. Little else was known about the feeding ecology of these whales, however. Field biologists like John Calambokidis knew the whales had three favourite areas—the Gulf of the Farallones, Monterey Bay and around the Channel Islands—but it was unclear *why* these areas had such abundant food. And it wasn't like you could steer your boat to the same GPS coordinates day after day and reliably find blue whales. "It's important for people to understand that blue whales aren't going to a single place where there's always food," says Don Croll, who specializes in ecology and evolutionary biology at the University of California Santa Cruz. "It's a dynamic system."

In the mid-1990s, Croll set out to learn more about these dynamics. Up to that point, Calambokidis's decade of work had been focused on individual whales. Croll and his long-time collaborator, Bernie Tershy, took a different tack. "We started looking at the foraging ecology of whales from the prey's perspective, as opposed to the whale's perspective. Because blue whales feed only on krill, which graze directly on phytoplankton, it's a simple system compared with most other large mammals that feed on a variety of things at different times of the year. We tried to understand where the food is, when it's there, and why it's there. Then we just plugged whales into that." The task that Croll, Tershy and their colleagues set for themselves was more difficult than it may sound—there is a reason no one had a comprehensive picture of the blue whale's summer feeding habitat off California. To obtain one, you would need simultaneously to measure the concentrations of phytoplankton and krill in many different areas to see where prey was most abundant, keep that up over several years so you could note any seasonal peaks, and figure out the role of environmental factors, including winds, currents, salinity, light levels, temperature and seafloor topography.

At the most basic level, Croll explains, the productive waters off California are the result of a system called coastal upwelling. When ocean animals and plants die, some of their remains sink and decay, so deeper water is richer in nutrients such as phosphates, nitrates and iron. Water near the surface, on the other hand, is largely devoid of this marine compost, but it is exposed to sunlight, the other essential component of photosynthesis. That's why upwelling is a fundamental part of any food-rich area in the ocean: when cool water is pushed to the surface, the two key ingredients mix, and the combination of sunlight and nutrients leads to the growth of phytoplankton.

Several forces can create upwelling. In some areas it is caused by tides, while elsewhere it results from deep currents running into a seamount and being steered toward the surface. Coastal upwelling systems are more complex and more unusual, since a number of factors must combine to draw up the cool water. First, you need a coastline along the eastern boundary of an ocean and, in the northern hemisphere, a prevailing wind from the north that creates a current. In this case, it's the cool, southbound California Current, which snakes its way down the Pacific coast, 60 to 120 miles. The Coriolis effect, caused by the west-to-east rotation of the earth, nudges the current to the right of the wind's direction, which drags surface water away from the coast and forces a colder layer up to take its place. "One misperception people have is that blue whales are feeding right in the California Current, but that's not what they're doing," Croll says. "The upwelling isn't occurring offshore—the coldest water is actually right next to shore." Coastal upwelling is intensified wherever points of land jut out. "Wind velocity is intensified as it goes around headlands, and therefore more cold water is brought to the surface, so you have a peak in nutrient availability."

To find a good blue whale feeding habitat, then, you start by looking for cool water making its way to the surface. But measuring sea-surface temperatures, which can be done using satellites in real time, is only the beginning. "The biology has to happen," Croll says. "Blue whales don't eat cold, nutrient-rich water." The next step is finding the phytoplankton blooms, and researchers now understand that this doesn't simply mean looking where the upwelling is most intense. "It's going to be somewhat removed. The phytoplankton is going to be downstream from the peak of upwelling, because the cold water that comes to the surface doesn't just stay there, it continues to move. As it moves away, you get a phytoplankton community developing." This community can develop in a hurry—phytoplankton blooms can double in size during a single day.

Krill, of course, do not appear and multiply quite that quickly. Some portion of the adult population overwinters each year, and when the peak of upwelling occurs in February and March, these animals emerge to spawn and then die. "Once that happens, it takes two to three months for the larvae that are produced from that spawning event to reach maturity," Croll explains, which means that the year's first generation, or cohort, reaches adulthood by May or June. "Then, if there's a lot of good upwelling, that cohort will spawn and have another cohort. And if it's a really good year, there'll be a third one. The ball will roll for two or three spawning events, and then once the upwelling slows down, things peter out and a handful of adults will overwinter again. That's why you can't just say, OK, there's been upwelling for the last three weeks, so krill should be there." While some animals eat krill larvae, blue whales do not seem to—they will seek out areas where they can expect to find their prey fully grown.

Continuing to follow these leads, Croll and his team surveyed the waters off California to get an understanding of the oceanographic conditions that lead to large concentrations of blue whale food. Here again, the scientists were able to discern a trend: the krill tend to gather in areas where there is a steep increase in depth, such as a break in the continental shelf or a submarine canyon. The reason is that in daylight krill need to descend to deep water—usually 300 feet or more—to avoid predators, which they do in tight formation. At night, they come back to the surface and spread out to feed under the cover of darkness. By convening near an area with steep topography, then, they achieve a balance: they make use of the deep water during the day, while still remaining close enough to shore to take advantage of the plankton-rich, recently upwelled water at night. The irony is that the two methods krill use to reduce their chances of being eaten by birds and fish—their nightly plunge and their habit of schooling together—have precisely the opposite effect when it comes to whales. After all, what suits a hungry blue whale better than a dense cloud of krill 300 feet below the surface? "Those defensive behaviours actually make them whale food," Croll says. "But my guess is that whales aren't the dominant predator that these things have to worry about, because a lot of animals eat krill. Whales aren't driving the system, they're riding the system—they're the tail, not the dog."

When Croll and Tershy fit all of these pieces together, an explanation emerged for why blue whales repeatedly visit three principal areas off California: these locations are all downstream from headlands, which cause more intense upwelling, and they all have topography steep enough to allow krill to make their daily migrations. Upstream from the Gulf of the Farallones, for example, is the fingerlike promontory of Point Reyes, and immediately west of the Farallon Islands is an undersea escarpment. Monterey Bay is bracketed by Point Ano Nuevo and Point Sur and its seafloor is bisected by a huge canyon. Finally, Point Conception leans out over the Santa Barbara Channel, which is right on the continental shelf. "Not all of these upwelling areas are going on at the same time," Croll stresses. "You've got differences in local winds, which means that sometimes the Gulf of the Farallones is turned on, and sometimes the area south of Point Sur is turned on, and sometimes the Channel Islands are turned on. The whales move between these areas, checking to see where the most krill is."

That's a key point: they are searching for *the most* krill. The areas blue whales prefer are not the only places where krill is found along the west coast, but Croll's research has underlined an important fact of blue whale feeding habits: they cannot afford to waste energy snacking on smatterings of krill if the promise of something better is around the corner. This is the downside of enormous body size: their prey requirements are so large that blues need very hearty meals, and they must constantly weigh the relative merits of staying put and eating what is available versus packing up to go in search of more. This interpretation meshes nicely with what Bruce Mate's satellite tags have shown: blue whales stopping in one area, then zipping off for a while, only to return again later. "We can all appreciate the difference between looking at a smorgasbord and a bowl of rice," Mate says. "They know when they're being successful enough, and if they aren't, I think they just press on. Blue whales don't sit around and wait very often—it's not a good strategy for them." Whether the whales remember the spots where they are likely to find the densest krill swarms from one year to the next, or whether they simply roam around until they happen upon them, is a nagging question. "I believe they know these traditional feeding areas, these places where there is a lot of wind-driven upwelling, but I don't think they know, if they're going to the Farallones, for example, that krill is going to be abundant when they arrive. They might get there and be disappointed, so they just keep heading up the line. We'll see blue whales in the south end of the Santa Barbara Channel, and then the winds will die off and in three or four days the animals will get antsy. Some of them will stay, but some of them will be at the Farallones three days later. Moving 100 or 150 miles a day is feasible for this animal."

This, at last, hits on a couple of the reasons why blue whales evolved their gargantuan size. For any animal, the relative energy cost of travelling declines as body size increases. So, at normal swimming speed, a blue whale expends, in terms of calories per pound of body weight, less energy than a dolphin or smaller whale. For a 75-foot blue, a mile is just 70 body lengths, and a hundred miles is an easy day's jaunt. "Their spatial scale is so far beyond ours, that it's difficult for us to consider what distance means to them," Croll says. They can browse between areas of the ocean with minimal effort, making it possible for them to exploit the ocean's spotty resources. Along the way, their enormous fat stores—blubber makes up about a quarter of their body mass—allow them to subsist during the inevitable periods when they are searching unsuccessfully for food. Since large body size is a buffer against the patchiness of krill swarms, it is easy to see how natural selection would favour the burliest individuals.[4]

No one knows precisely how many daily calories it takes to power a blue whale, but even a ballpark figure gives some idea of the scope of the animal's energy needs. Dale Rice estimated that a blue whale around 80 tons would require an average of 1.5 million calories a day—the highest prey demand of any animal. Since they feed most voraciously in summer in order to store as much energy as possible, blue whales may take in 3 million calories on a good day. In the eastern North Pacific, where the prey favoured by blues is less than an inch long and it takes about 300 to make an ounce, that works out to some 4 tons of food and 40 million individual krill. It should be pointed out that this is a maximum figure. Photographs taken on whaling ships, where blues were opened up after being hauled from their Antarctic feeding grounds, show piles of stomach contents that look like they were poured from a dump truck. To imply, as so many popular sources do, that blue whales eat several tons of food every day is no more accurate than suggesting Christmas dinner is a typical human meal.

To take in that much prey, blues have evolved some highly specialized feeding anatomy. The first is their baleen, which is made of keratin, a protein similar to the material in horse hooves and human fingernails. A blue whale has about 300 to 350 baleen plates hanging from the roof of its mouth, each less than a quarter of an inch thick and shaped like a knife blade. While the baleen of a bowhead whale can be over 10 feet long, rorquals have much wider, coarser and shorter plates, with the lengthiest reaching only about 3 feet. The side that faces the inside of the animal's mouth is fringed with hairlike bristles that form a dense mat. It is here that millions of krill make their futile death throes before entering the dark tunnel of the whale's esophagus. Surprisingly, given the giant scale of almost everything else in a blue whale's anatomy, its gullet is actually rather small; according to one account from the whaling era, "a four-pound loaf would probably choke the largest Rorqual."[5]

Baleen whales feed in quite different ways. Right whales and bowheads are skim feeders, cruising along with their mouths open and straining zooplankton from a continuous stream of water. Grey whales, by contrast, do not need to be moving in order to force water into their mouths; they can draw it in with suction, which they often do while rooting in the seabed for invertebrates. Blue whales and other rorquals, however, are primarily lunge feeders, or gulpers. They take in their prey and the surrounding seawater one enormous mouthful at a time, then close their jaws almost completely and expel the water, trapping the food in their baleen. Maximizing the amount they can engulf requires a unique adaptation: from directly under the "chin" to the umbilicus, or about half the animal's length, stretches a series of expandable pleats forming a pouch. This ventral pouch is made from two kinds of tissue: a grooved and particularly elastic layer of blubber covering an equally extensible layer of muscle. When a blue whale opens its lower jaw while swimming or lunging forward, the force of the water causes the pouch to expand. The lower jaw drops almost 90 degrees and the ventral pouch blows out like a bullfrog's, doubling the whale's already enormous diameter. One researcher estimated that a blue whale feeding this way can engulf 1,000 tons of prey-laden water at a time, which is absurd, but serves as a reminder that even scientists can get giddy when observing the spectacle. A more reasonable (but still astounding) estimate is about 15,000 gallons of seawater. That's a 65-ton mouthful. A blue whale's feeding lunge is, in the words of one biologist, "the largest biomechanical action in the animal kingdom."[6]

One of the most bizarre and mysterious parts of a blue whale's feeding anatomy is its tongue. At birth, a rorqual's tongue is a solid, muscular structure that the baby whale uses to suckle. But after the animal is weaned, the muscle fibre gives way to more elastic tissue and the organ becomes a flaccid, deformable mass. Whalers often remarked on its jelly-like consistency, and wrote of how it would slip and slide dangerously on the deck of a ship, threatening to give unlucky flensers a fatal tongue-lashing.

Scientists aboard whaling ships examined and described the oral structures in baleen whales, but little was known about how these actually worked when the animal was feeding. In the early 1980s,

American biologist Richard Lambertsen took a closer look at rorqual feeding mechanisms, examining dozens of fin and sei whales as they were being processed by Icelandic whalers. He confirmed the surprising elasticity of the tongue, noting that when the flensers tugged it with a winch, the organ stretched to more than twice its normal length before springing back when cut free. When a dead minke washed ashore in Maine in March 1981, Lambertsen and his colleagues used a more innovative research method: they removed its head and suspended it from two chains so they could examine the movement of the tongue as the head was tilted and rotated. Lambertsen's examination was the first close-up observation of how the rorqual tongue changed shape when the angle was altered. Most remarkably, when the head was tipped backward 40 to 50 degrees, the tongue slipped under the esophagus and larynx and turned itself inside out, lining the floor of the mouth. Imagine putting your hand into a surgical glove and pulling it back out—a rorqual's tongue apparently undergoes something similar, transforming itself into a hollow sac. Lambertsen suggested this occurs whenever the whale lowers its jaw to feed: "The force required to transform the tongue would come from the flow of water into the open mouth generated by the animal's rapid forward locomotion."[7] This action would in turn cause the ventral pouch to expand, enlarging the capacity of the whale's mouth. But despite the natural stretchiness of the tongue and the ventral pouch, Lambertsen felt that their elastic recoil would not be enough to forcibly expel the water through the whale's baleen. Some additional force, muscular or otherwise, must also be lending a hand.

A decade later, Lambertsen was still mulling over how blues and other rorquals were able to pull up their lower jaw after engulfing a mouthful of seawater and krill. This question may seem mundane to a human being who has never given much thought to the energy required to chew, but it's important to remember that a blue whale lunging at 15 miles per hour with its mouth open creates an astounding amount of hydrodynamic drag. Picture yourself riding in a car at that speed with the top half of your body sticking out of the sunroof. Then imagine unfurling a parachute and trying to hang on to it. A blue whale must overcome a similar force every time it closes its mouth to trap its prey before they escape.

Lambertsen's dissections revealed that one of the anatomical tools they use to do this is a previously unknown structure called the frontomandibular stay. This is a thick, fibrous appendage of the temporalis muscle, one of several muscles used in chewing. Lambertsen found that a dead rorqual's jaw would naturally fall open to an angle of about 70 degrees. Beyond that there was a good deal of resistance—he used a steam winch to pry the jaw wider during his investigations—and by 85 to 90 degrees the jaw muscles were under extreme tension. When the winch was released, the jaw sprang back on its own. The newly discovered piece of anatomy, he discovered, worked as both a brake and a spring. From the time the mouth begins to open until it reaches about 70 degrees, the frontomandibular stay has no effect at all. However, between 70 and 90 degrees, it begins to impart resistance. Then, when the gaping jaw reaches 90 degrees, "an enormously powerful, high amplitude impulse would suddenly be delivered" to completely stop the jaw from opening wider. At this point, the momentum caused by the whale's forward motion is recaptured and put to work pulling the jaw closed. Once that is accomplished, the elastic blubber and the muscles in the ventral pouch do the rest, squeezing the water and krill toward the baleen plates. "What is avoided in this way," Lambertsen wrote, "is the completely dysfunctional situation in which each huge gulp taken by the whale promptly bounces back out of its mouth, prey included."[8]

There is still one critical part of the blue whale's feeding technique that has long been unclear. Once the whale expels the seawater from its mouth, how exactly does it remove the prey trapped in its baleen? Alex Werth, a biologist at Hampden-Sydney College in Virginia, makes the analogy of a dip net used to skim a swimming pool. The pool cleaner might remove the debris from the net by scraping with his hand, shaking the net to free the material, or spraying water through the net from the underside. Werth proposes three similar techniques used by baleen whales: removal by scraping with the tongue, head shaking to jostle the prey loose and "hydrodynamic flushing," or backwashing water into the mouth. All three may be used to some degree by baleen whales, but in rorquals, removal with the tongue seems the most likely method. They probably scrape their prey from the back of the plates, channel it into a mass by lifting both sides of the tongue, then swallow it, though no one really knows how this is accomplished. As Werth readily admits, "a clear, Jonah's-eye view inside a live whale's mouth is unattainable."9

The problem—the same one that crops up relentlessly when studying almost any aspect of blue whales' lives—is that observing feeding animals is extremely difficult. Because krill migrate well below the surface at night, where they congregate into convenient meal-sized swarms, blue whales do most of their feeding at depth. Blues do occasionally feed at the surface, however, where they offer brief glimpses of their distended ventral pouches, and this is where Brian Kot saw an opportunity. Kot, a graduate student at the University of California Los Angeles, joined Richard Sears's team in Quebec and soon learned that the St. Lawrence is *the* place to see surface-feeding blue whales. "The oceanography of this area is so strong that it just pushes the krill right up to the surface, sometimes regardless of the light levels. So we can have lunge-feeding blue whales at noon. Very bizarre, very unusual." But also tantalizing, since it offered the possibility of finally being able to collect hard data about the whales' feeding behaviour and biomechanics. "I'm going to one of the best feeding grounds in the world for these whales, and I'm getting digital video of surface feeding. I want to analyze how these animals are using their feeding anatomy while it's in plain sight—no speculation about how it's being done. I'm not measuring a bloated animal that's been on a beach for two weeks."

Since 2004, Kot has been shooting video of surface-feeding blue whales, minkes and fin whales and analyzing the four different lunging techniques these rorquals use. The first is the vertical lunge, a spectacular display common in humpbacks and sometimes used by minkes, where the whale comes straight up out of the water with its mouth open. "You don't see that with blue whales much," Kot says. "Richard says he's seen it in the past, but I never have." Then there is the oblique lunge, common in minkes but unknown in blues, where the animal emerges from the water at about 45 degrees and then does a "chin flop." A third technique, called a lateral lunge, has the whale rolling onto its left or right side with its mouth open. Finally, there's the ventral lunge, where the grooved underside of the whale comes up first. "Sometimes that's all you see breaking the surface—you don't see the jaw, you just see this big belly." The latter two techniques—the lateral and ventral lunges—are by far the most commonly observed in blue whales at the surface.

Why would the whales use different techniques when their prey is always the same? "There could be a couple factors. Sometimes the upwelling pushes the krill onto the shelf near shore, into just 20 metres of water depth, and a blue whale is longer than that—they would actually be able to touch the bottom and the surface at the same time. That might affect whether they can actually get upside down and lunge." Prey density might be another factor, Kot adds. "For blue whales, when you only see the ventral pouch breaking the surface, it seems to be a less intense lunge. There's this arcing and rolling, where just the pouch comes up, then they slowly roll upright and you see the pectoral fin, then the eye, and then they take a breath. It's a very slow process. Other times the jaw comes way out of the water—imagine this 3- or 4-metre-tall mandible trucking along the surface next to your boat. It's a spectacular event. When you see that, you get the feeling that these animals are excited. I'm thinking that when there's more krill near the surface, they might be lunging with more intensity—perhaps intentionally trapping the krill up against the surface and engulfing them before they disperse."

Kot stresses that there isn't enough data to prove this yet. "But these animals are lunging for a reason—they're trying to catch their prey in the most effective and efficient way possible. If the krill are amassed just below the surface, versus being spread less densely from the surface down to 10 or 15 metres, that could be the factor that determines which technique they use. I don't know that yet, but sometimes out here the water is red with krill right below the surface—it's *so* thick—and that's when I see this high-intensity lunging, with the jaw way out of the water. Whereas on the depth sounder you might have 14 metres of krill, but it looks rather sparse, and we just see the slow arcing and rolling." Kot says that the most intense lunging comes when the weather is cloudy with a slight drizzle—conditions that tempt field researchers to catch up on their paperwork. "We can sometimes be good-weather biologists—we like to go out when it's sunny. But I'm sort of the opposite. When it's overcast and dark and there's a little bit of rain, I fire up my boat and go out. It's a little uncomfortable, and my camera can get a little wet, but that's the best time to see surface feeding, because that's when the krill come to the surface."

Kot's other main interest is what happens to the water inside the ventral pouch during feeding. How fast is the water travelling when it enters the mouth, and how fast is it being forced back through the baleen? To study this, he uses an innovative technique: digital video that can be broken down frame by frame. By examining the frames, he can see the crest of the wave that ripples along the outside of the pouch after the whale engulfs a mouthful of seawater. "When the water comes in, it balls up at the back of the pouch, creates a big wave, and then that wave travels forward toward the baleen plates. I've quantified how fast that wave is moving." Using software to look at his video one frame at a time, Kot places a digital landmark (a simple dot) on the moving wave in each frame. He then measures the change in distance between the landmark and a fixed reference point, such as the rostrum or a pectoral fin, during a sequence of frames. Because he knows his video contains 30 frames per second, he can divide the distance by the time to determine the velocity of the water inside the whale's mouth. For example, if the wave travels 2 feet over five consecutive frames, its velocity is about 16 feet per second.

When comparing his findings among blues, minkes and fin whales, Kot was surprised to find that the speed of the water was fastest inside the blue whales' mouths. "That's counterintuitive, because minkes are very acrobatic and blues are relatively slow, so you would think that the animals lunging faster would have a faster rebounding flow out of the ventral pouch. In blues it's about 16 feet per second, in finbacks it's about 13 feet per second, and with minkes it's less than 7. And I thought, that doesn't make sense." Then Kot noticed that the differences in speed were directly related to differences in the body size of the three species. In other words, the water in a blue whale's mouth moves about 2.5 times faster than in a minke's because it has to travel about 2.5 times farther. "The flow is actually coming back out in a similar fashion. Whether it's a blue whale or a minke whale, they're both using their feeding anatomy in a similar way. That's pretty amazing to me." It's worth stressing the difference in magnitude, however. Lambertsen measured the capacity of the ventral pouch in his severed minke head at almost 160 gallons. Kot estimates that the force of this volume of water moving through a minke's mouth would be less than 500 newtons. (A 2-pound block resting on the ground exerts a force of just under 10 newtons.) No one has ever measured the capacity of a blue whale's ventral pouch, but Kot used a conservative estimate of 13,000 gallons. That amount of water moving at 16 feet per second would create a biomechanical tsunami with a force he estimates at over 90,000 newtons.

As his work continues, Kot has found that the speed at which water is expelled from the ventral pouch varies even among whales of the same species. "This may suggest that they're not only engulfing large mouthfuls of prey, but are sometimes taking smaller quantities as well. Richard Sears says he's seen minkes feed on only a few fish at a time—sort of like snacking. And I've seen blues take short-duration, small gulps at the surface—15 or 20 seconds, which is much shorter than the more typical full engulfment technique, which can take about 60 seconds. These smaller gulps seem to be more lateral lunges, and the larger gulps are more likely to be inverted, ventral lunges." As Don Croll, Bruce Mate and others have emphasized, the food requirements of a blue whale are so high

that they normally seek out and exploit only the densest aggrega-
tions of krill. But perhaps by varying their behaviour—downshifting to a less energetic feeding mode when prey is sparse—they have
come up with a way of tiding themselves over until the pickings
improve.

The detailed story of how gigantic blue whales subsist on tiny
krill—which is at the heart of why they have evolved into the planet's
largest creatures—is still unfolding. Kot jokes that he's come up
with an idea for the ultimate experiment. "I'm going to go to Radio
Shack and build a 'krillcam.' I'll build a hundred of them, maybe a
thousand. They'd be really small, maybe an inch long. And when I
have lunging blue whales around I'm going to throw them out and
hope that one of them gets engulfed." Who knows, the krillcam
may be just the thing to capture that elusive Jonah's-eye view.

TWELVE

THE FUTURE IS BLUE

THE JACKET COPY of George Small's 1971 book, *The Blue Whale*, begins with a chilling assertion: "The blue whale is not extinct yet. But a tragic history of slaughter, ignorance and indifference has led to the unhappy conclusion that it soon will be." Even 18 years later, a *New York Times* article announced that "the majestic blue whale, the biggest animal ever to live on earth, appears far closer to extinction than scientists have believed." It went on to cite a "devastating" study that found only 453 blue whales in the Antarctic, "where they expected to find 10 times that many."[1] Bringing up that same study in 1991, Richard Ellis wrote: "The greatest creature in the history of the planet may become extinct in our lifetime."[2] What should we make of these doomsday predictions today? Is the blue whale still at risk of extinction?

There is no question that the species has failed to recover from the slaughter it endured, even a half-century after the killing stopped. The International Whaling Commission banned the hunting of blue whales in the North Atlantic in 1955, and throughout the world a decade later. Thousands of blues were later illegally killed in the Indian Ocean, but even these crimes had ceased by the early 1970s. One would expect blue whales to be showing signs of bouncing back after so many years of complete protection, and yet they are still on the endangered species lists of the United States, Canada, Australia and Chile. As recently as 2002, Canada officially separated its

Pacific and Atlantic populations and declared both endangered. That same year, a scientific committee in Australia announced that the blue whale "is likely to become extinct in nature in New South Wales unless the circumstances and factors threatening its survival or evolutionary development cease to operate." None of this sounds any more encouraging than the bleak assessments of decades past.

But while saying so may raise the ire of the Chicken Littles, the fact is that many of these pessimistic assessments are misleading. The Canadian government's report, for example, explained its decision about the Pacific population by pointing out that "the rarity of sightings (both visual and acoustic) suggests their numbers are currently very low (significantly less than 250 mature individuals)."[3] This is true for Canadian waters, and following the country's criteria for protecting species, North Pacific blue whales clearly fall under the heading of endangered. But given the recent discovery that B.C. waters form just a small part of the range of a large and thriving population that moves between Alaska and Costa Rica, one has to consider the context. United States law considers all blue whales endangered, too, but that assessment was made in 1970, when sightings were extraordinarily rare. The situation has certainly changed enough to warrant another look. It will probably get one within a decade, and it will be difficult to argue that eastern North Pacific blue whales are endangered.

For blues elsewhere the situation is quite different. This is a key point: when assessing the status of the species, each population needs to be considered individually. The World Conservation Union's Red List of Threatened Animals, for example, considers the eastern North Pacific stock as "lower risk/conservation dependent," the North Atlantic population as vulnerable, and only the Antarctic blues as endangered. (Pygmies, which may make up half of all blue whales left today, have not been assessed because of the lack of data.) However, the Red List also classifies blue whales as globally endangered, a label it defines as "facing a very high risk of extinction in the wild in the near future." This blanket assessment is "based on an estimated decline of at least 50% in worldwide total abundance over the last three generations, assuming a generation time of roughly 20–25 years."[4] That worldwide decline is indisputable—indeed, it was more than 90 percent—but it is misleading when you acknowledge that 330,000 of the blue whales killed, or more than 87 percent, were in the Antarctic. Every other blue whale population was depleted by whaling, surely, but to nowhere near the same extent, as the Red List's experts admit in the small print. Even if all the northern hemisphere and pygmy blue whales were to recover to pre-whaling numbers, the worldwide abundance would still remain well below 50 percent. Would the species still be considered globally endangered?

This is not to imply that the Pacific, Atlantic and Indian Oceans are teeming with blue whales and that we need not give a second thought to their conservation—far from it. It simply means that not all of the populations can be lumped together if words like "endangered" are to have any meaning. In the Antarctic, the population is still *less than 1 percent* of its pre-whaling total. This makes it one of the most endangered populations of whales on the globe, and it needs the highest level of protection. The challenges faced by other blues should be kept in perspective. The entire North Pacific, west and east, may have held only 5,000 blues before whaling, and the North Atlantic population may have been not much larger. In the southern hemisphere, no one knows how numerous pygmy blue whales were before they were identified in the 1960s, nor are there any reliable estimates of present numbers. Any honest prognosis, therefore, should not include gloomy rhetoric or claims, like the one made by the scientists in New South Wales, that they are at risk of going extinct, even locally, without our immediate intervention. A more rational approach is to continue studying these animals, to learn more about their relationship to the environment, and to ensure that humans are not interfering with that relationship. And, above all, to make certain that no country is ever again allowed to hunt blue whales for any purpose.

Before looking at the future of blue whales in more depth, it helps to get some perspective from other species that were heavily hunted. Sperm whales were slaughtered for more than two centuries, from the first Yankee whalers until the IWC's total ban in 1986. Some populations still face significant threats, but today sperm whales are not endangered in any ocean, and the global population is probably more than 400,000. There are several reasons why this species has fared better than most. For one, whalers targeted the much-larger males and killed a smaller number of breeding females. (The situation was reversed for blues—both sexes were targeted, but no living creature was larger and more valuable to whalers than a sexually mature female blue whale.) Sperm whales also inhabit the deep ocean, away from most harmful human activity, and prey largely on species of squid that are not threatened by overfishing.

Several baleen whale populations have also recovered well after being protected, though others have not. Eastern Pacific grey whales were hunted until they were virtually extinct, but after a total ban in 1938 they have rebounded to more than 21,000, close to their pre-whaling levels. Here again, the whales' life history—specifically, their predictable near-shore migration route and their loyalty to certain breeding areas—made them easy to protect. The population of grey whales in the western Pacific, by contrast, is just over a hundred animals and remains critically endangered. Humpbacks, too, have proven resilient after their terrible overexploitation, and are increasing in all areas where they are being monitored. Finbacks, close cousins of the blues, died in staggering numbers: more than 700,000 in the Antarctic alone. Today, they are abundant in the northern hemisphere, while in the Southern Ocean only a tiny fraction of the original population remains. Bowheads in the Arctic, the quarry of European whalers for centuries, are recovering overall, though some local populations remain critical. Southern hemisphere right whales are thriving, while those in the North Atlantic—protected since the 1930s—number only about 400 and still face serious threats. The bottom line is that whale populations typically increase after they are protected by hunting, but the rate varies greatly depending on the species' behaviour and biology, and how directly they are affected by human activity.

Each of the world's blue whale populations faces a different suite of potential threats, and the jury is still out on the degree to which any of them poses a significant danger. Monitoring any increase in their numbers is also difficult for all the reasons that are now familiar: the global population structure is not well understood; pre-whaling estimates are too sketchy to provide a baseline; they are distributed over very wide and often remote areas; and their unpredictable movements make them difficult to monitor over long periods. Blues were long thought to have been extirpated in the Gulf of Alaska, for example, but it now seems clear that they have simply changed their travel patterns. Richard Sears watched blue whales vanish from the Mingan Islands in Quebec in the early 1990s; no one knows the reason for the disappearance, but the whales are probably alive and well in some other part of the western North Atlantic. When blues are no longer seen in an area where they were once common, it does not mean they have fallen off the face of the earth.

One factor working in the blue whale's favour is longevity; they may live for 50 years, or even much longer, with few known natural causes of death. A very small number may get trapped in wind- or current-driven ice in the North Atlantic in early spring—at least 28 have died along the southern shore of Newfoundland, but these records date back to 1868, which works out to just one every five years. This does not appear to be a problem for other populations; the Antarctic is the only other region where it would be possible, and blues are not known to venture into the pack ice. The only other natural threat is attacks by orcas, especially on young animals. Early in the whaling era, seamen often reported that even the mighty blue could be preyed on by groups of killer whales. The yarns tended to sound like this one:

Two killers can dispose of a 150-ton leviathan. Each sinks his long, sharp teeth into the monster's jaw, one on each side, then together they pull his head under water. The orca can hold its breath for as long as twenty minutes, the big blue only about five. Because of this difference, the deadly dolphins can keep their gigantic prey submerged until he drowns. Then they slash open his mouth and snap off his enormous tongue, letting the rest of the carcass float away untouched. Even their teeth will not penetrate the tough, inch-thick hide.[5]

Given the nonsense that appears in this paragraph—blues can hold their breath for longer than 20 minutes, and their skin is far less than an inch thick and not particularly tough—it would be easy to write off the whole story as yet another myth. But this one is true. A crew of scientists off Baja stumbled upon just such a spectacle in 1979, photographing and filming an attack for the first time. About 30 orcas descended upon a young blue whale, ripped off its dorsal fin, shredded its flukes and tore enormous chunks of flesh from its body, the largest estimated to be 6 feet square. One of the orcas did indeed try to keep the unfortunate victim from surfacing to breathe. The attack lasted at least five hours, after which the blue whale was still alive, through mortally wounded. Richard Sears has found that up to a quarter of the blue whales ID'd in Baja bear scars from orcas, suggesting that attacks are common, though rarely successful.

Although harpoons are no longer aimed at blues, humans do still kill a small number of them inadvertently. The animals in the North Atlantic, North Pacific and possibly other areas are occasionally struck and killed by ships. In March 1998, a tanker entered Narragansett Bay, Rhode Island, and the crew of a pilot boat spotted the body of a large whale impaled on its bow. It turned out to be a juvenile blue, which was probably struck and killed off Nova Scotia and carried into port unnoticed. (The skeleton, with fibreglass casts replacing the smashed vertebrae, is now mounted in the New Bedford Whaling Museum.) In the St. Lawrence about one-sixth of all the blue whales in Sears's catalogue have scars or deep wounds caused by collisions with vessels, including one with an amputated fluke. Every year, about 5,000 large ships pass by Les Escoumins— a blue whale hot spot—and a 30-knot ferry began zipping between the north and south shore in 1997. While no blue whale fatality from these vessels has ever been confirmed, Sears points out that "if it's a big 300-foot ship, it would be like us getting hit by a Mack truck. The whale would probably get opened up and fall to the bottom and we'd never know about it."

The only region with a significant number of blues killed by ships is the eastern North Pacific, though from 1980 until 2006 no more than half a dozen were reported. Then, during a terrible three-week stretch in September 2007, three blue whales were found dead in the Santa Barbara Channel, all with smashed bones that indicated they were the victims of ship strikes. While it is possible that some factor helped disorient the whales—military sonar and demoic acid from toxic algae blooms were mentioned but never proven—the likely reason for the carnage was simply that the channel, a heavily travelled shipping lane, was filled with blue whales that summer because krill was more abundant than ever. Collisions seem to occur with neither the whale nor the ship's crew being aware of each other until it is too late, which suggests that even very large vessels may be surprisingly quiet when they are travelling toward the animal. Peter Gill and Margie Morrice, whose study area off southern Australia is also a busy shipping lane, once observed a blue whale making evasive manoeuvres only after the ship bearing down on it was less than 1,500 feet away.

Other than shipping, human activity has not caused serious problems for blues today. Becoming tangled in fishing gear, a major cause of death for North Atlantic right whales, is much less of an issue for the massive blues. Richard Sears notes that many blue whales in the St. Lawrence have superficial scarring that may come from nets or hooks, but only three have died from drowning since 1979. A potentially bigger worry is the growing number of whale-watchers in places where blues come close to shore. In the U.S. and

Canada, where blues are protected as endangered species, whale-watch boats are supposed to keep back 100 yards and avoid following any animal for prolonged periods. New regulations drawn up for the blue whales off Chile, an increasingly popular tourist attraction, will keep boats back 300 yards. Not every operator respects these rules, and the sheer number of boats on the water may harass the whales. Sears has seen more than half a dozen converging on a single blue whale in the St. Lawrence, while Monterey Bay can be so crowded with whale-watchers that scientists find it impractical to work there in high season. The optimist might point out that only 25 years ago there were no blue whales off California for either researchers or tourists to experience.

On his first day in Iceland in July 1996, Richard Sears set out from the small fishing village of Grindavik and encountered almost 50 killer whales, some minkes and a few humpbacks. Those sightings alone would have made for a dream day, but what sticks most firmly in Sears's memory is that he encountered two blue whale calves swimming alongside their mothers. Spotting a blue whale calf off Iceland in summer is not that unusual, but what was jarring for Sears is that he found these cow-calf pairs so quickly. He has spent a hundred days in the St. Lawrence every year, yet he has seen just 17 calves there. Now here were two on his first day in the eastern North Atlantic. And there have been more since: in only about 40 days of effort spread over 10 seasons, Sears and his team have observed about the same number of offspring in Iceland as they have in the St. Lawrence in almost three decades.

Fertility is obviously important to the recovery of any species, and mothers arriving on the feeding grounds with calves is an encouraging sign. Eastern North Pacific blue whales seem to be breeding regularly: Sears and Calambokidis have encountered many calves in Baja, and they are not uncommon off California. Chile is positively bursting with calves—10 were spotted in 2006 alone. And, as Sears discovered that first day in Iceland, those in the eastern North Atlantic are thriving, too. In fact, this population appears to be increasing and may number more than 1,000. Sears isn't sure whether the western group is growing. (The standard mark-recapture technique does not work in the St. Lawrence because the samples are not random—some animals reliably return there year after year, while about a third have been seen only once.) But the decided lack of calves has left him wondering why other blue whales seem to be reproducing at a much greater rate than the ones in eastern Canada.

He has considered several possible explanations. One is that the mothers heading north in spring simply choose to stay outside the Gulf. "That would be surprising," he says. If mothers and calves were not coming into the St. Lawrence, the animals Sears encounters would be overwhelmingly male, but that's not the case. "We have many mature females here. It's actually about a 50–50 sex ratio—there are a few more males, but not a huge difference." Another explanation is that pollutants in the St. Lawrence might be interfering with the whales' reproduction. The estuary's famous belugas have long been known to contain high levels of contaminants, and they consistently bear fewer offspring than their counterparts in the Arctic Ocean. Is it possible that blue whales, too, are being harmed by toxins in the water?

The north shore of the Gulf is not densely populated—fewer than 400,000 people live there, with no town larger than 40,000—and agriculture and heavy industry are limited. However, the region is downstream from the Great Lakes and the St. Lawrence River, both of which support enormous urban populations. The Gulf is also fed by the Saguenay River basin, home to several pulp and paper mills and aluminum smelters. Add this to ever-present airborne pollutants and the result is a marine ecosystem with worrisome levels of pollution. Both the Gulf and the estuary contain an alphabet soup of PCBs, DDT, HCHs and other contaminants, many of which were ingredients in pesticides. Although these substances are now banned or controlled, some take a very long time to break down, and significant amounts are still found in the atmosphere and in sediments. As animals ingest them, the toxins accumulate in fat cells, getting more and more concentrated as they move up the food chain from phytoplankton to krill to fish to whale. Blubber-bearing marine mammals can be storehouses of the stuff, though by the late 1980s no one really knew the contaminant loads of baleen whales in the St. Lawrence. Sears decided to find out.

Since 1988, MICS researchers have been taking biopsies of ror-quals in the St. Lawrence and have sent many of their blubber samples to Trent University in Peterborough, Ontario, where they are analyzed for dozens of persistent contaminants. Early results showed that blue whales had higher concentrations of DDT than humpbacks, finbacks and minkes, although the sample size was small. The scientists who did the analysis suggested that blue whales may be less efficient than other rorquals in breaking down this contaminant. In a later study that compared humpbacks and blues, the two species showed similar overall concentrations of PCBs, but blue whales had higher levels of certain highly chlorinated varieties. This was unexpected because humpbacks eat fish, while blues dine only on krill. The higher you go on the food chain—and fish are at least one rung above krill—the more likely you are to have high concentrations of persistent compounds, so one would expect humpbacks to contain more, not equal amounts, and certainly not less. Finally, the study also demonstrated that humpback calves were carrying similar contaminant loads to their mothers, evidence that persistent pollutants are passed along when they nurse. There were no blue whale calves among the sample—the lack of baby blues prompted the study in the first place—but it is not a great leap to assume that female blues offload contaminants in the same way.

So are blue whales in the St. Lawrence having fewer calves because pollution is interfering with their reproduction? Certainly these contaminants have caused major problems in other species. DDT was devastating to nesting falcons and eagles who ingested it:

their eggs became so soft that most broke before hatching. Studies on mink, once commonly ranched in the Great Lakes basin, showed that PCBs can severely impede the ability of females to bear young. But the concentrations of these compounds in St. Lawrence blues are far lower than those that wreaked havoc on other animals. Even their neighbours, the belugas, carry loads at least a hundred times greater. However, Sears has biopsied many blue whales in Icelandic waters and has found that this fertile population is carrying far lighter contaminant loads. So it is possible that even small amounts may be disruptive to blues. For what it's worth, the contaminants do not seem to be affecting the reproduction of other rorquals in the St. Lawrence. "The humpbacks are pumping out 15 calves a year just in Mingan, and we had 11 finback calves in 2005," Sears says. "Maybe the humpbacks and finbacks we're seeing here don't go as high up in the estuary, so they're not feeding where the water is most polluted. Maybe the fatty globules in krill hold PCBs more—we don't know yet. There hasn't been much analysis of the krill in the St. Lawrence." Sears would love to see this done, but the Canadian government has shown no interest in paying for it. "I called up one of their toxicologists and he said, 'They don't want to know, they don't want any more bad news.' They've already had enough with the belugas."

The recovery of blue whales, not just in the North Atlantic but everywhere, may be threatened by two other factors, though these also need further study. The first is climate change, which could disrupt their most important feeding areas. The good news is that blue whales are able to find food-rich upwelling areas even when these move from year to year, or even week to week, as environmental factors change, and satellite tags have repeatedly showed just how well adapted they are for travelling great distances to forage. When their prey declined in the Gulf of Alaska, the eastern North Pacific blue whales apparently just shifted their summer feeding grounds farther south. But no one knows how an overall rise in sea-surface temperatures might affect krill populations worldwide. The average air temperature over the Southern Ocean in winter has risen more than 2°C during the last 30 years, and the sea ice is retreating in some regions. Less sea ice can lead to less plankton, and in some parts of the Antarctic scientists have noted an 80 percent drop in krill over the same three-decade time frame. However, in other regions the sea ice is showing the opposite trend. Climate change is an ongoing concern for all species, including *Homo sapiens*, but what threat it may pose to blue whales is currently a very large unknown.

The second potentially worrisome trend is the increasing amount of noise in the ocean. Both military and commercial vessels now routinely conduct experiments that emit loud, low-frequency sound pulses designed to locate submarines or explore for undersea oil and gas. Most sea creatures are unlikely to be disturbed by these sounds, but baleen whales are an exception. Since blue and fin whales make very low vocalizations, their ears are almost certainly attuned to these frequencies, and many scientists worry that loud and persistent human-made sounds will harm them. There is some dramatic evidence that mid-frequency naval sonar may actually kill beaked whales—three mass strandings have been linked to military tests since 1995.

To see how blue and fin whales react to low-frequency sounds, Don Croll, Chris Clark, John Calambokidis and others performed a five-week experiment off San Nicolas Island in California in 1997. The U.S. Navy gave them access to the sonar system on one of its vessels, and the researchers carefully monitored the whales in the area to see whether they changed their feeding behaviour and vocalizations in response to the sounds. They also tried to identify as many individuals as possible to determine whether the same ones were returning, or if they were learning to avoid the area. Their results showed no obvious responses to the noise—the whales' movements were tied far more closely to where the krill was. While this is encouraging, Croll and his colleagues point out that the prolonged effect of low-frequency sound may be cumulative, like the buildup of greenhouse gases in the atmosphere. If blue whales do communicate over very long distances in order to find mates, the increased difficulty of doing that in a noisy ocean could affect their recovery. The image that so galvanized environmentalists in the 1970s—the lonely blue whale calling out in an empty ocean—needs to be revised. It may be more a case of many blue whales calling, but none of them able to hear the others for all the background noise.

Just how many blues are left in the Southern Ocean after the carnage of the whaling era has been a source of controversy since the Committee of Three estimated in 1963 that only 600 remained. A decade later the whales probably numbered between 150 and 840. Since the late 1970s, researchers have completed three sets of surveys of the waters around the Antarctic continent, each of which took several years. While none of these was specifically designed to count blue whales, the spotters did report all the species they sighted, and scientists have used these data to estimate the number of surviving blues. The initial results were grim. The first round of surveys, completed between 1978 and 1984, returned with a figure of 453. This was the number that appeared in the *New York Times* and other articles declaring that blues were on the brink of extinction. A second group of surveys between 1985 and 1991 estimated 559 blue whales—hardly cause to break out the champagne, though it was 23 percent higher than the previous figure. The third series was still not complete in 2001, but the preliminary estimate came in at 1,100. This was almost double the previous figure, though most scientific papers continued to state that Antarctic blue whales were showing no signs of recovery despite decades of complete protection.

It is important to stress that these head counts have wide margins of error. As with all line-transect surveys, the population estimates are extrapolated from the number of actual sightings—no individual cruise before 2001 spotted more than seven blue whales during any year. This scarcity made the statistical models less reliable than they were for more common species, such as minkes. But if the number of Antarctic blue whales really had grown from a couple of hundred individuals to well over a thousand, that was an encouraging trend. While pursuing his PhD at the University of Washington, Trevor Branch set out to see whether this apparent increase was real. He started by gathering all of the SOWER data and combining it with Japanese data from as far back as 1968. Then he considered biological factors such as when blues reach sexual maturity, how long they live, how often they bear calves and the ratio of females to males. (None of these were known for certain, and had to be assigned a range of values.) He also looked at the rate of increase that other baleen whale populations were showing, to see what was possible. When he was done, his model showed that Antarctic blues were increasing by between 1.4 and 11.6 percent a year, with the most likely rate being 7.3 percent. The population in 1996 would have been about 1,700, he argued, significantly more than anyone had estimated before.

When Branch presented his initial findings at the IWC meeting in Berlin in 2003, the scientists charged with assessing Antarctic blues were not convinced. Many of them felt that such an increase was biologically impossible, given that blue whales are thought to have no more than one calf every two or three years, and that the only other blue whale population known to be increasing is the one off Iceland—and even that increase, reported to be 4.9 percent, has been disputed. The following year, Branch presented a revised version of his model that addressed these objections, and this time the scientists "agreed that this research represents a considerable advance on previous work." Some continue to question the 7.3 percent figure, and Branch freely admits the rate may be lower—or higher, for that matter—but says the important point is that the whales are certainly increasing, not plummeting toward extinction. A follow-up paper in 2007 confirmed the increase and announced that the most recent series of surveys, which covered 99.7 percent of the ice-free water below 60°S, saw more blue whales than ever before and estimated the population at 2,280.

One might think this news would be met with a chorus of cheers from conservationists and scientists, but the reception was rather cool. Part of the reason may simply have been that some questioned the methodology—Branch's models are extremely complex and can be opaque even to people who are schooled in statistics. But that was also true of the survey methods used to demonstrate that Antarctic blue whales were heading toward extinction, and these were all over the news in 1989. A recovering population, by contrast, does not seem to generate the same reaction—as journalists like to say, no front-page headline ever reads "Plane lands safely." One reviewer who rejected Branch's paper for the prominent journal *Nature* even commented, "If this result were reported for a fungus instead of a whale, I doubt it would be publishable."

There is a political element to all of this. Environmental groups and research scientists are wary about splashing too much good news on the front page. They rely on funding from governments and the public, and it is much easier to draw attention to your work when you are trying to protect critically endangered animals. A slow recovery rate is good, since it is evidence that your intervention is succeeding, but if there is too big of an increase—and 7 percent is quite large for a whale population—your message sounds less urgent: "No need to worry about those blue whales in the Antarctic any more. This study shows that they're thriving." In this case, however, that kind of complacency is wholly inappropriate. Imagine that you invested $240,000 in the New York Stock Exchange just before the crash of 1929. Overnight your portfolio is reduced to a mere $360. But instead of throwing yourself from a window in despair, you invest what is left, averaging returns of 7 percent per year. At that rate, you might just beat the market, but after 10 years, your account would still be just $619. After 50 years, your portfolio would be all of $9,262. At what point do you start feeling enthusiastic about your recovery? By this same measure, even if Antarctic blue whales continue to increase by 7 percent annually—and that is an optimistic "if"—they would not approach their pre-whaling numbers until 2068.

There is one final danger to overemphasizing the recovery of blue whales, and this involves the internal politics of the IWC. Back in 1972, the United States tabled the first proposal for a moratorium on commercial whaling at the commission's annual meeting in London. At that time there were 14 member countries and only four voted in favour; since the motion required a three-quarters majority, it was soundly defeated. For the next decade, similar proposals were made every year and they always failed to get the required number of votes. After 1979, however, the IWC underwent a rapid transformation. Its membership ballooned with 23 new countries, many recruited by the U.S. and environmental groups with the goal of tipping the balance in favour of nations that would support the moratorium. It worked: at the 1982 meeting in Brighton, England, the motion passed with 25 votes in favour and just 7 against, with 5 abstentions, and commercial whaling ended in 1986. (Several countries that voted for the ban, including Jamaica, Belize, the Philippines and Egypt, left the IWC shortly thereafter, lending credence to the claims that they had been brought in as hired guns.) In just a few years, the IWC went from a group of nations with a vested interest in commercial whaling to an organization dominated by non-whaling countries that were committed to killing the industry. The Japanese, with support from Norway, Russia, and Iceland, which rejoined the IWC in 2002 after a 10-year absence, have continually lobbied for an end to the moratorium and a return to the commercial harvest of species they feel can sustain it, notably minke whales. Year after year, their attempts have been thwarted. But early in the new century the pendulum began to swing the other way. In 2002, 7 more nations joined the IWC, followed by 21 others over the next five years, and almost none have any history of whaling. Environmental groups regularly accuse the Japanese of buying votes from developing countries with aid money. By 2007, the IWC had 77 members, divided more or less down the middle. Neither side can produce the necessary three-quarters majority, but the Japanese-led camp currently has the momentum.

The Japanese have relied on two main arguments when lobbying for a return to commercial whaling. The first is that whales eat so much that they threaten the livelihood of countries that rely on the fishing industry. Whenever they are accused of buying votes at the IWC, the Japanese counter by arguing that developing nations support whaling because they "worry that if the whale population continues to grow unchecked, fish stocks could for all intents and purposes dry up."[6] Japanese scientists regularly claim that whales eat up 300 to 500 million tons of fish each year—three to six times more than humans. This argument, even if it were not based on bogus numbers, has little relevance in the Southern Ocean, since Antarctic baleen whales feed almost entirely on krill, which has little commercial value. But the Japanese have another argument for this region: there is so much competition for krill that endangered whales do not stand a chance: "What is impeding blue whales from recovery?" writes Masayuki Komatsu, an official in his country's Ministry of Agriculture, Forestry and Fisheries. "It is believed that minke whales, and not humans, are the culprits."[7]

For years, the Japanese have argued that there are 760,000 minkes in the Antarctic, far more than during the heyday of commercial whaling. The IWC's Scientific Committee accepted this as the best available estimate in 1990, though many recently have questioned its reliability, and later surveys suggested that minkes have since declined by as much as 65 percent. While such a precipitous drop is unlikely—there may have been problems with the survey methodology—the Scientific Committee acknowledged in 2001 that it simply could not estimate the minke population with any precision. Yet Japanese whaling interests persist in invoking the 760,000 figure and argue that minkes can easily support a controlled hunt. (Setting aside the ethical questions, their current practice of killing several hundred minkes annually does not threaten the species in the Antarctic.) Indeed, Komatsu contends that such a hunt is integral to the blue whale's future:

> If we want to allow blue whales to recover to [their] original level, it would be necessary to cull a considerable number of minke whales. This is a law of nature with which mankind has interfered. Since mankind has broken the law and skewed the balance of nature, it is a duty imposed upon us to act responsibly and bring it back to the proper balance.[8]

Many scientists point out that this argument ignores the fact that the Antarctic ecosystem supported vastly larger whale populations in the past, so there is no reason to expect competition to be an issue now. In any event, it is naive to think that the Japanese concern for blue whales is altruistic. Seiji Ohsumi, a long-time delegate at the IWC, has made no secret of what will follow if and when blues are again numerous: "We are looking forward to the day when this valuable resource [recovers] to an optimum level and their rational use resumes."[9] An unfortunate choice of words given that nothing about blue whale hunting was ever rational.

It would be alarmist to suggest that any nation is close to resuming a commercial hunt on Antarctic blue whales. The chances of that occurring in the foreseeable future are remote. But the Japanese may soon broach the idea of killing blue whales for "research." Since 1987, the year after the moratorium, Japan has run a program of so-called scientific whaling, during which they kill several hundred whales annually. The IWC allows member countries to issue special permits for this activity, so the Japanese are doing nothing illegal, but their program has long been criticized on both ethical and scientific grounds. While no one argues that examining specimens during the whaling era led to enormous gains in understanding whale biology, few scientists outside Japan see any value in continuing to kill whales for research. Indeed, despite 20 years and more than 10,000 dead whales, the Japanese research program has produced a relatively small number of papers in peer-reviewed journals. (By contrast, the Japanese contributions to non-lethal whale studies, such as the SOWER surveys in the Antarctic, have been invaluable.) The research program—which is highly subsidized by the Japanese government—is primarily a way for Japan to keep its foot in the door until it successfully ushers in a return to commercial whaling.

Whales killed under scientific permit may be processed and sold for food—this is allowed by the IWC as a way to fund the research—and it is still possible to buy whale meat in Asian markets. Not all of this meat is from legitimate catches, however. Before 2000, Japanese research focused almost exclusively on minkes, yet in the 1990s a group of plainclothes scientists went into Japanese and South Korean markets to buy whale meat and used DNA analysis to prove that only about half came from minkes. They found the genetic fingerprints of at least two dozen fin whales, a humpback, a blue whale and a blue/fin hybrid—the DNA detectives learned that this last animal was killed by Icelandic whalers in 1989. These findings were a smoking gun that shattered any claims that scientific whaling is a well-regulated activity that targets only plentiful species.

One of the most worrisome trends in scientific whaling is that it is expanding rapidly to include other species that have only recently begun to recover. Since 2000, the Japanese have broadened their dubious research to include 5 to 10 sperm whales annually, then 50 Bryde's whales, then 100 sei whales, all in addition to their usual take of more than 500 minkes. In 2005, the Japanese doubled the number of Antarctic minkes killed and added 10 fin whales, for a one-season death toll of 1,243 whales. If that wasn't enough, they announced that in 2007 they would launch a new program that would kill 50 fin whales and 50 humpbacks in the first season alone. One of their stated aims of this long-term research is to focus on "minke, humpback and fin whales and *possibly other species* in the Antarctic ecosystem that are major predators of Antarctic krill"[10] (emphasis added). In Iceland, meanwhile, whalers killed their first fin whale since the moratorium in October 2006, arguing that the

North Atlantic stocks had recovered to the point where a controlled hunt was sustainable.

It is hard to see these events as anything other than a slippery slope, and they help explain why more scientists and conservationists are not trumpeting from the rooftops about every increase in blue whale populations. The Japanese, meanwhile, rarely miss an opportunity to emphasize the recovery of some whale species, even citing numbers that most scientists reject as unrealistic; the government of Japan has even claimed that humpback and fin whales are recovering by close to 30 percent, while the IWC's Scientific Committee concludes that it is biologically impossible for these species to increase more than about 10.6 percent. No one has yet announced plans to take blue whales, but it would be foolish to discount the possibility of that occurring in the future, especially if they are shown to be swelling in numbers. If one thing was crystal clear throughout the bloody history of whaling it is that the industry is incapable of regulating itself. Blue whales, the most valuable species of them all, may never be able to rest easy.

The blue whale's greatest enemy may be human ignorance. Our understanding of blue whales has increased dramatically since the 1980s, but many secrets remain. Scientists need to know more about how blue whales find food if they are to understand how climate change might affect them. They need to learn more about the whales' complex and unpredictable migrations if they are to protect the most important habitats. They are closer than they were a century ago to understanding where blues go in winter, but this mystery may never be solved—and perhaps that is for the best. The blue whale, the largest animal ever to inhabit our planet, owes our species nothing. Humans, on the other hand, who almost destroyed these most marvellous of creatures, are not yet finished paying reparations. We should continue to learn about their lives, and to protect them from the harpoon with our laws. Above all, however, we should leave them in peace.

Author's Note

Wild Blue was a three-year journey, and like any long trip it included incredible memories, difficult obstacles and an opportunity for new friendships.

I will never be able to repay the debt I owe to Richard Sears. After 30 years of working with blue whales, he might have felt protective of these animals, and resentful of an outsider who thought he could write about them. Instead, Richard spent countless hours sharing his experience with me. (And to thank him, I brought rain every time I visited.) So much of what Richard has learned over the decades has not been written down. I hope this book goes some small way toward sharing his wisdom with others.

John Calambokidis not only tolerated my presence during my first ride-along, during which I was seasick on his boat, but even invited me to spend a week on the water with his team a month later. During our many conversations since then, he was refreshingly open in discussing his findings. He, too, expressed a desire to share his vast experience with the public. There is no question that I could not have written this book without his participation.

Trevor Branch of the University of Washington was extraordinarily generous with his time and encouragement. He patiently explained his own work to me, and shared many hard-to-find papers from his impressive blue whale library. Trevor read the entire manuscript with his critical eye, correcting me when I got something wrong and offering kind words when I got it right.

James Meiklejohn, founder of the Salvesen's Ex-Whaler's Club, helped me sort through the fact and fiction in the whaling literature, and was such a diligent fact-checker that he visited the Whaling Museum in Sandefjord to personally measure a harpoon so I could describe it accurately. Through many phone calls and faxes, he shared his incredible adventures in the Antarctic, even while suffering ill health. He kindly arranged my interview with Hans Bechmann, who is now in his late nineties, a privilege that few other journalists have enjoyed. I am grateful to both Mr. Bechmann and Alf Mathisen for sharing their stories of life as a whale gunner.

Gibbie Fraser is a wonderful storyteller who made me feel like I was on board the whale catcher with him. He and his wife, Laurena, also gathered and scanned several photographs from the Antarctic whaling era, for which I am deeply thankful. For readers who would like to learn more about Fraser and the men who went south with him, I strongly recommend his self-published book, *Shetland's Whalers Remember*.

Thomas Doniol-Valcroze helped me track down many scientific papers, including a few exotic sources I never expected to find. He read sections of the manuscript and offered his insights despite being extremely busy with his own PhD work and his young family.

Kate Stafford, Catherine Berchok and Erin Oleson patiently answered my questions about their work, supplied me with recordings and helped me understand the world of blue whale acoustics. William Cummings, now retired from the U.S. Navy, also generously reviewed part of the chapter on acoustics, supplied reprints of decades-old papers and clarified the role of early researchers in this field.

Brian Kot's patience and enthusiasm, as well as his extraordinary video, helped me understand the feeding habits of blue whales. I have no doubt that he will eventually be the first to build a working krillcam.

Thank you also to those who extended their hospitality during my travels. Andrea Bendlin and the MICS team welcomed me into their home away from home in Quebec, and made sure I was well fed and entertained. Lee Bradford, captain of the *John H. Martin*, invited me to his beautiful home for a meal and good company while I was travelling alone in California.

I would like to thank all of the other experts who shared their time and expertise, including Abigail Alling, Ed Cassano, Phil Clapham, Don Croll, Carole Conway, Bárbara Galletti, Peter Gill, Jim Gilpatrick, Rodrigo Hucke-Gaete, Steven Katona, Benjamin Kahn, Rick LeDuc, Bruce Mate, Mark McDonald, Chris Metcalfe, Margie Morrice, Jan Erik Ringstad, Alex Werth and Terrie Williams. I hope they feel I have represented their work accurately.

I am also grateful to John Bryant, a Melville scholar at Hofstra University in Hempstead, New York, for sharing his insights on the blue whale as it appears in *Moby-Dick*.

Don Sedgwick and Shaun Bradley of Transatlantic Literary Agency not only found a home for my idea, but showed a genuine interest in the project. Janice Zawerbny, Patrick Crean and everyone at Thomas Allen Publishers were a pleasure to work with, even after I missed their deadlines, while Liba Berry and Wendy Thomas lent their editorial expertise to clean up the manuscript.

Finally, as always, thank you to Wendy, Jaimie and Erick for listening to three years' worth of whale stories at the dinner table.

Notes

Important sources are listed here by lead author and date of publication. For full citations, see Sources.

PREFACE

1 Chadwick 2006, 208.
2 Hoyt 2005a, 29.

ONE: INTO THE BLUE

1 R. Payne, Seeing Blues Whales, *Voyage of the Odyssey*, March 12, 2002 (http://www.pbs.org/odyssey/odyssey/20020312_log_transcript.html).
2 Olsen 1914–15.
3 Deraniyagala 1948.
4 Barnes 1996, 69.
5 A. Junaidi, The legend of "lamafa," *Jakarta Post*, June 12, 2007.
6 quoted in True 1899.
7 Dewhurst 1834, xiii.
8 Stein 1992.
9 Conover 1996.
10 Melville 1851, 236.
11 Ibid., 240–41.

12 Lane 1991.

13 quoted in True 1904. The original article is G.O. Sars, Om "Blaavalen" (*Balaenoptera sibbaldi*, Gray), med Bemærkninger om nogle andre ved Finmarkens Kyster forekommende Hvaldyr. *Christiana Vidensk-Selsk Forhandl* 1874: 227–41.

Other key sources: Aguilar and Lockyer 1987, Ash 1964, Bérubé and Aguilar 1998, Best 1998, Branch et al. 2004, Calambokidis and Steiger 1997, Carwardine 1999, Chadwick 2006, Ellis 1991, Hucke-Gaete et al. 2004, George et al. 1999, Laurie 1937, Mackintosh and Wheeler 1929, Matthews 1978, Ohsumi 1979, Olsen and Sunde 2002, Ommanney 1971, Rice 1986, Sears 2002, Severin 1999, Slijper 1962, Small 1971, Spilliaert et al. 1991, Vincent 1949, Wilson and Reeder 2005.

TWO: THE GREATEST WAR

1 Jackson 1978, 160.

2 Scammon 1874, 72.

3 Ellis 1991, xi.

4 Wheeler 2004, 27.

5 Francis 1990, 186.

6 Tønnessen and Johnsen 1982, 151.

7 Ibid., 1982, 160.

8 Ellis 1991, 354.

9 Ibid., 1991, 346.

10 C. Wilson, *The History of Unilever*, quoted in Jackson 1978, 184.

11 Francis 1990, 195.

12 Villiers 1931, 97–98.

13 Ibid., 1931, 102.

14 Murphy 1947, 207.

15 Villiers 1931, 124–25.

16 Ibid., 125–26.

Other key sources: Bennett 1932, Branch et al. 2004, Brandt 1940, Cook and Wisner 1973, Euller 1970, Mackintosh and Wheeler 1929, Ommanney 1971, Reeves et al. 1998, Small 1971.

THREE: BACK TO THE ICE

1 Fraser 2001, 2.
2 Ash 1964, 68.
3 McLaughlin 1962, 102.
4 Fraser 2001, 6.
5 McLaughlin 1962, 108.
6 Small 1971, 102.
7 Ash 1964, 21.
8 Small 1971, 107.
9 Ibid., 115, 142.

Other key sources: Branch et al. 2004, Clark 1973a, Clark 1973b, Ellis 1991, Euller 1970, Francis 1990, Jackson 1978, Tønnessen and Johnsen 1982.

FOUR: ABOUT-FACE

1 Mowat 1972, 59–60.
2 Ommanney 1971, 10.
3 Discovery Committee 1937, 15.
4 review of the *Report on the Progress of the Discovery Committee's Investigations* in *Geographical Journal* 90(3): 280–81 (September 1937).
5 Mackintosh 1950.
6 Villiers 1931.
7 Ash 1964, 123.
8 Ellis 1991, xii.
9 Small 1971, 192–93.
10 McHugh 1974, 335.
11 Small 1971, vii.
12 all quotes in paragraph from Villiers 1931.
13 Ash 1964, 19.
14 Ibid., 32.
15 Ibid., 125.
16 Ibid., 119.
17 McLaughlin 1962, 18.

18 Katona et al. 1993, 66.
19 Ellis 1991, 436.
20 Ellis 1986, 14.
21 Smith 1979.
22 Small 1971, 205.
23 Ibid., 204.

Other key sources: Birnie 1985, Branch et al. 2004, Branch et al. 2007a, Branch et al. 2007b, Brownell and Yablokov 2002, Cherfas 1998, Francis 1990, Freeman 1990, Jackson 1978, Mackintosh 1929, Mackintosh 1965, Payne 1995, Rayner 1940, Tønnessen and Johnsen 1982.

FIVE: ST. LAWRENCE BLUES

1 Rice 1986, 41.

Other key sources: Branch, Stafford et al. 2007, Chadwick 2006, Christensen 1955, Corkeron and Connor 1999, Donovan 1991, Edds and Macfarlane 1987, Hoyt 2005, Jonsgård 1955, Jonsgård 1966, Lavigne et al. 1990, Mitchell 1974, Ramp et al. 2006, Sigurjónsson and Gunnlaugsson 1990, Sears et al. 1990, Sears et al. 1999, Sears and Calambokidis 2002, Sears and Larsen 2002, Tønnessen and Johnsen 1982, Wenzel et al. 1998.

SIX: WEST COAST BLUES

1 Scammon 1874, 73.
2 Rice 1966, 11.
3 Rice 1974, 179.
4 Donahue et al. 1997.

Other key sources: Barlow 1995, Branch et al. 2007b, Burtenshaw et al. 2004, Calambokidis et al. 1990, Calambokidis and Barlow 2004, Calambokidis and Barlow in press, Calambokidis and Steiger 1997, Donovan 1991, Mackintosh and Wheeler 1929, Reilly and Thayer 1990, Reeves et al. 1998, Reeves et al. 2003, Rice 1986, Russell 2001, Sears and Calambokidis 2002, Yochem and Leatherwood 1985.

SEVEN: BLUES' CLUES

1 Chadwick 2006, 212.

Other key sources: Carwardine 1999, Costa 1993, Acevedo-Gutiérrez et al. 2002, Calambokidis and Steiger 1997, Calambokidis 2002, Calambokidis 2003, Etnoyer et al. 2006, Goldbogen et al. 2006, Mate et al. 1999, Mate et al. 2007.

EIGHT: BLUE NOTES

1 Chadwick 2006, 217.

2 Payne and Webb 1971.

3 Ibid.

4 Ibid.

5 Payne 1995, 185.

6 Ibid., 180.

7 *Blue Whales '93*, compact disc, Cornell University Laboratory of Ornithology, Bioacoustics Research Program (notes by Chris Clark).

8 Clark 1994.

9 Payne 1995, 180.

10 Stafford et al. 1999a.

11 quoted in Dan Joling, Rare blue whales spotted in Alaska: 30 years since last confirmed sighting, Associated Press, July 28, 2004.

12 see McDonald et al. 2006.

Other key sources: Aroyan et al. 2002, Beamish and Mitchell 1971, Berchok et al. 2006, Carwardine 1991, Clark 1995, Cummings and Thompson 1971, Rankin et al. 2006, Rivers 1997, McDonald et al. 1995, Mellinger and Clark 2003, Oleson et al. 2007a, Oleson et al. 2007b, Stafford et al. 1998, Stafford et al. 1999b, Stafford 2003, Stafford et al. 2004, Stafford and Moore 2005, Thompson et al. 1996, Watkins et al. 2004, Wiggins et al. 2005.

NINE: THE 79-FOOT PYGMY AND OTHER STORIES

1 Pliny, *Naturalis Historia*, Book 11, Chapter 62, translation by John Bostock.

2 Aristotle, *The History of Animals*, Book III.

3 Pliny, *Naturalis Historia*, Book 9, Chapter 88 (62).

4 Small 1971, 21.

5 Scoresby 1820.

6 Blyth 1859.

7 Burmeister 1872.

8 Ichihara 1961.

9 Ichihara, 1966.

10 Small 1971, 200, 202.

11 Mowat 1972, 43.

12 Rice 1998.

13 Aguayo 1974, 212.

14 quoted in Jennifer Jacquet, Chile: Proposing a Haven for Blue Whales, *AquaNews* (Vancouver Aquarium), June 3, 2005.

15 Hucke-Gaete et al. 2004.

Other key sources: Berzin 1978, Branch, Stafford et al. 2007, Cabrera et al. 2006, Carwardine 1999, Ellis 1991, Findlay et al. 1998, Galletti Vernazzani et al. 2006, Galletti Vernazzani et al. 2007, Gill 2002, Gill and Morrice 2003, Hoyt 2005b, Hucke-Gaete et al. 2004, Hucke-Gaete et al. 2005, Hucke-Gaete and Mate 2005, Kahn and Pet 2002, Kahn 2005, Kahn et al. 2007, Ljungblad et al. 1997, Mackintosh 1963, Ruud 1956, Scammon 1874, Tønnessen and Johnsen 1982, True 1899, Zemsky and Boronin 1964.

TEN: BLUE GENES

1 Branch 2006, 5.

2 Branch et al. 2007b, 22.

Other key sources: Alling and Payne 1987, Best 1993, Branch 2007, Branch et al. 2007a, Conway 2005, Donovan 2000, Kato et al. 2002, LeDuc et al. in press, Ljungblad et al. 1998, Mackintosh 1963, Palacios 1999, Sigurjónsson and Gunnlaugsson 1990, Small 1971, Stafford et al. 1999a, Stafford et al. 2005a.

ELEVEN: WHY SIZE MATTERS

1 Rice 1986.
2 Ibid., 1986.
3 Chadwick 2006, 208.
4 see Croll and Tershy 2002.
5 Bennett 1932, 42.
6 P. Brodie, Noise generated by the jaw actions of feeding fin whales. *Canadian Journal of Zoology* 71 (1983): 2546–50, quoted in Acevedo-Gutiérrez et al. 2002.
7 Lambertsen 1983.
8 Lambertsen et al. 1995.
9 Werth 2001.

Other key sources: Arnold et al. 2005, Branch et al. 2007b, Calambokidis and Steiger 1997, Chadwick 2001, Chadwick 2006, Croll et al. 2005, Croll 1998, Everson 2000, Gingerich 2005, Mackintosh and Wheeler 1929, Ommanney 1971, Reeves et al. 1998, Sears 2002, Werth 2000, Werth 2001, Wong 2003, Yochem and Leatherwood 1985. I am indebted to Trevor Branch and Jim Gilpatrick for providing unpublished data that were integral to this chapter.

TWELVE: THE FUTURE IS BLUE

1 Stevens 1989.
2 Ellis 1991, 21.
3 Sears and Calambokidis 2002.
4 Reeves et al. 2003.
5 Henry 1950, 199–200.
6 Komatsu 2002b.
7 Komatsu 2002a, 30.
8 Komatsu 2002a, 30–31.
9 Ohsumi 1994.
10 Government of Japan 2005.

Other key sources: Baker et al. 2000, Best 1993, Branch 2007, Branch and Butterworth 2001, Cabrera et al. 2007, Cipriano and Baker 1999,

Clapham et al. 1999, Clapham et al. 2007, Clapham and Brownell 1996, Croll et al. 2001, Croll et al. 2002, Freeman 1990, Gales et al. 2005, Gauthier et al. 1997, Gill and Morrice 2003, Hammond et al. 1990, Metcalfe et al. 2004, Ramp et al. 2006, Reeves et al. 1998, Sigurjónsson and Gunnlaugsson 1990, Stenson et al. 2003, Tarpy 1979.

Sources

BOOKS

Ash, Christopher. 1964. *Whaler's Eye*. London: George Allen & Unwin.

Barnes, R.H. 1996. *Sea Hunters of Indonesia: Fishers and Weavers of Lamalera*, Oxford: Clarendon Press.

Bennett, A.G. 1932. *Whaling in the Antarctic*. New York: Henry Holt and Company.

Birnie, Patricia. 1985. *International Regulation of Whaling*. New York: Oceana.

Brandt, Karl. 1940. *Whale Oil: An Economic Analysis*. Palo Alto, CA: Stanford University Press.

Calambokidis, John, and Gretchen Steiger. 1997. *Blue Whales*. Stillwater, MN: Voyageur Press.

Carwardine, Mark, ed. 1999. *Whales, Dolphins and Porpoises*. 2nd ed. New York: Checkmark Books.

Chadwick, Douglas H. 2006. *The Grandest of Lives: Eye to Eye With Whales*. San Francisco: Sierra Club.

Cherfas, Jeremy. 1988. *The Hunting of the Whale*. London: The Bodley Head.

Chrisp, J. 1958. *South of Cape Horn: A Story of Antarctic Whaling*. London: Robert Hale Limited.

Cook, A.S. 1921. *The Old English Physiologus*. New Haven: Yale University Press.

Cook, Joseph J., and William L. Wisner. 1973. *Blue Whale: Vanishing Leviathan*. New York: Dodd, Mead and Company.

Deput, Steve. 2003. *The Barnsley Whale*. Edinburgh and London: Mainstream Publishing.

Dewhurst, Henry William. 1834. *The Natural History of the Order Cetacea and the Oceanic Inhabitants of the Arctic Regions*. London.

Ellis, Richard. 1991. *Men and Whales*. New York: Alfred A. Knopf.

Euller, John. 1970. *Whaling World*. Garden City, NY: Doubleday and Company.

Everson, Inigo. 2001. *Krill: Biology, Ecology and Fisheries*. Oxford: Blackwell Science.

Francis, Daniel. 1990. *A History of World Whaling*. Markham, ON: Viking.

Fraser, Gibbie. 2001. *Shetland's Whalers Remember*. Self-published.

Henry, Thomas R. 1950. *The Story of Antarctica*. New York: William Sloane Associates.

Hoyt, Erich. 2005a. *Whale Rescue*. Richmond Hill, ON: Firefly Books.

———. 2005b. *Marine Protected Areas for Whales, Dolphins and Porpoises*. London: Earthscan.

Jackson, Gordon. 1978. *The British Whaling Trade*. London: Adam & Charles Black.

Katona, Steven K., Valerie Rough and David T. Richardson. 1993. *Field Guide to Whales, Porpoises, and Seals from Cape Cod to Newfoundland*. 4th revised ed. Washington: Smithsonian Institution.

Komatsu, M. 2002. *The Truth Behind the Whaling Dispute*. Japan: Institute of Cetacean Research.

Lien, Jon, and Steven Katona. 1990. *A Guide to the Photographic Identification of Individual Whales*. St. John's, NF: Breakwater Books.

Mackintosh, N.A. 1965. *The Stocks of Whales*. London: Fishing News (Books) Ltd.

McLaughlin, W.R.D. 1962. *Call to the South: A Story of British Whaling in the Antarctic*. London: George Harrap and Co.

Matthews, L. Harrison. 1978. *The Natural History of the Whale*, London: Weidenfeld and Nicholson.

Melville, Herman. 1851. *Moby-Dick; or The Whale*. Reprint, London and New York: Penguin. 1972.

Mowat, Farley. 1972. *A Whale for the Killing*. Reprint, Toronto: Seal Books (McClelland-Bantam). 1988.

Murphy, Robert Cushman. 1947. *Logbook for Grace*. Reprint, New York: Time Inc. 1965.

Ommanney, F.D. 1971. *Lost Leviathan: Whales and Whaling*. New York: Dodd, Mead & Company.

Payne, Roger. 1995. *Among Whales*. New York: Scribner.

Rice, D.W. 1998. *Marine Mammals of the World. Systematics and Distribution*.

Special Publication Number 4. The Society for Marine Mammalogy. Lawrence, KN: Allen Press.

Robertson, R.B. 1954. *Of Whales and Men*. New York: Alfred A. Knopf.

Russell, Dick. 2001. *The Eye of the Whale*. New York: Simon & Schuster.

Scammon, Charles M. 1874. *The Marine Mammals of the North-western Coast of North America*. Reprint, Dover Publications, New York. 1968.

Scoresby, William. 1820. *An Account of the Arctic Regions: With a History and Description of the Northern Whale-fishery, Volume 1: The Arctic*. Reprint, Newton Abbot, UK: David & Charles. 1969.

Severin, Tim. 1999. *In Search of Moby Dick*. London: Little, Brown and Company.

Sibbald, Sir Robert. 1803. *The history, ancient and modern, of the sheriffdoms of Fife and Kinross, with a description of both, and of the firths of Forth and Tay, and the islands in them . . . with an account of the natural products of the land and waters*. London: R. Tullis.

Slijper, E.J. 1962. *Whales*. New York: Basic Books.

Small, George. 1971. *The Blue Whale*. New York and London: Columbia University Press.

Smith, Vincent. 1979. *The Last Blue Whale*. San Francisco: Harper & Row.

Tønnessen, J.N., and A.O. Johnsen. 1982. *A History of Modern Whaling*. Berkeley and Los Angeles: University of California Press.

True, Frederick W. 1904. *The Whalebone Whales of the Western North Atlantic*. Washington: Smithsonian Institution.

Villiers, A.J. 1931. *Whaling in the Frozen South*. Indianapolis: Bobbs-Merrill.

Vincent, Howard P. 1949. *The Trying-Out of Moby-Dick*. Boston: Houghton Mifflin.

Wheeler, Tony. 2004. *The Falklands & South Georgia Island*. Footscray, Australia: Lonely Planet.

Wilson D.E., and D.M. Reeder (eds.). 2005. *Mammal Species of the World*, 3rd ed. Baltimore: Johns Hopkins University Press.

SCIENTIFIC ARTICLES

Acevedo-Gutiérrez, A., D.A. Croll and B.R. Tershy. 2002. High feeding costs limit dive time in the largest whales. *Journal of Experimental Biology* 205: 1747–53.

Aguayo, A. 1974. Baleen whales off Chile, in *The Whale Problem*. Edited by W.E. Schevill. Cambridge, MA: Harvard University Press.

Aguilar, A., and C.H. Lockyer. 1987. Growth, physical maturity, and mortality of fin whales (*Balaenoptera physalus*) inhabiting the temperate waters of the northeast Atlantic. *Canadian Journal of Zoology* 65: 253–64.

Alling, A., and R. Payne. 1987. Songs of Indian Ocean blue whales, *Balaenoptera musculus*. Paper presented to the Meeting to Review the Indian Ocean Sanctuary, Seychelles, February 1987 (unpublished).

Arnold, P.W., et al. 2005. Gulping behaviour in rorqual whales: underwater observations and functional interpretation. *Memoirs of the Queensland Museum* 51: 309–32.

Aroyan J.L., et al. 2002. Acoustic models of sound production and propagation, in *Hearing by Whales and Dolphins*, edited by W.W.L. Au, A.N. Popper and R.R. Fay. New York: Springer-Verlag.

Baker, C.S., et al. 2000. Predicted decline of protected whales based on molecular genetic monitoring of Japanese and Korean markets. *Proceedings of the Royal Society of London*, Series B 267: 1191–99.

Beamish, P., and E. Mitchell. 1971. Ultrasonic sounds recorded in the presence of a blue whale *Balaenoptera musculus*. *Deep-Sea Research* 18: 803–06.

Berchok, C.L., D.L. Bradley and T.B. Gabrielson. 2006. St. Lawrence blue whale vocalizations revisited: characterization of calls detected from 1998 to 2001. *Journal of the Acoustical Society of America* 120: 2340–54.

Bérubé, M., and A. Aguilar. 1998. A new hybrid between a blue whale, *Balaenoptera musculus*, and a fin whale, *B. physalus*: frequency and implications of hybridization. *Marine Mammal Science* 14: 82–98.

Berzin, A.A. 1978. Whale distribution in tropical eastern Pacific waters. *Report of the International Whaling Commission* 28: 173–77.

Best, P. 1993. Increase rates in severely depleted stocks of baleen whales. *ICES Journal of Marine Science* 50: 169–86.

———. 1998. Blue whales off Namibia—a possible wintering ground for the Antarctic population. International Whaling Commission paper SC/50/CAW514.

———. 1989. Some Comments on the BIWS Catch Record Data Base. *Report of the International Whaling Commission* 39: 363–69.

Blyth, E. 1859. On the great rorqual of the Indian Ocean, with notices of other Cetals, and of the Syrenia or marine Pachyderma. *Journal of the Asiatic Society of Bengal* 28: 481–98.

Branch, T.A. 2006. Separating pygmy and Antarctic blue whales using ovarian corpora. International Whaling Commission paper SC/58/SH8.

———. 2007. Abundance of Antarctic blue whales south of 60°S from three complete circumpolar sets of surveys. International Whaling Commission paper SC/59/SH9.

Branch, T.A., and D.S. Butterworth. 2001. Estimates of abundance south of 60°S for cetacean species sighted frequently on the 1978/79 to 1997/98 IWC/SOWER sighting surveys. *Journal of Cetacean Research Management* 3: 251–70.

Branch, T.A., K. Matsuoka and T. Miyashita. 2004. Evidence for increases in Antarctic blue whales based on Bayesian modelling. *Marine Mammal Science* 20: 726–54.

Branch, T.A., et al. 2007a. Past and present distribution, densities and movements of blue whales *Balaenoptera musculus* in the southern hemisphere and northern Indian Ocean. *Mammal Review* 37: 116–75.

Branch, T.A., et al. 2007b. Separating southern blue whale subspecies based on length frequencies of sexually mature females. *Marine Mammal Science* 23: 803–33.

Brownell, R.L., and M.A. Donaghue. 1994. Southern hemisphere pelagic whaling for pygmy blue whales: review of catch statistics. International Whaling Commission paper SC/46/SH6.

Brownell, R.L., and A.V. Yablokov. 2002. Illegal and pirate whaling. In *Encyclopedia of Marine Mammals*. Edited by W.F. Perrin, B. Wursig and J.G.M. Thewissen. San Diego: Academic Press.

Burmeister, H. 1872. On *Balaenoptera patachonica* and *B. intermedia*. *Annals and Magazine of Natural History* 10 (fourth series): 413–18.

Burtenshaw, J.C., et al. 2004. Acoustic and satellite remote sensing of blue whale seasonality and habitat in the Northeast Pacific. *Deep-Sea Research* II 51: 967–86.

Cabrera, E., et al. 2006. Preliminary report on the photo-identification of blue whales off Isla de Chiloé, Chile, from 2004 to 2006. International Whaling Commission paper SC/58/ SH18.

Cabrera, E., B. Galletti Vernazzani and C.A. Carlson. 2007. Recommendations for whale watching guidelines in the blue whale feeding area of southern Chile. International Whaling Commission paper SC/59/WW15.

Calambokidis, J., et al. 1990. Sightings and movements of blue whales off central California 1986–88 from photo-identification of individuals. *Report of the International Whaling Commission*, Special Issue 12: 343–48.

Calambokidis, J., and J. Barlow. 2004. Abundance of blue and humpback whales in the eastern North Pacific estimated by capture-recapture and line-transect methods. *Marine Mammal Science* 20: 63–85.

Chapman, D.G. 1974. Status of Antarctic rorqual stocks. In *The Whale Problem*. Edited by W.E. Schevill. Cambridge, MA: Harvard University Press.

Christensen, G. 1955. The stocks of blue whales in the northern Atlantic. *Norsk Hvalfangst-Tidende* 44: 640–42.

Cipriano, F., and S.R. Palumbi. 1999. Genetic tracking of a protected whale. *Nature* 397: 307–8.

Clapham, P.J., and R.L. Brownell Jr. 1996. The potential for interspecific competition in baleen whales. *Report of the International Whaling Commission* 46: 361–67.

Clapham, P.J., S.B. Young and R.L. Brownell Jr. 1999. Baleen whales: conservation issues and the status of most endangered populations. *Mammal Review* 29: 35–60.

Clapham, P.J., et al. 2007. The whaling issue: conservation, confusion, and casuistry. *Marine Policy* 31: 314–19.

Clark, Christopher W. 1995. Applications of US Navy underwater hydrophone arrays for scientific research on whales. *Report of the International Whaling Commission* 45: 210–12.

Clark, Christopher W., and K.M. Fristrup. 1997. Whales '95: A combined visual and acoustic survey of blue and fin whales off southern California. Report of the International Whaling Commission 47: 583–600.

Clark, Colin W. 1973a. Profit maximization and the extinction of animal species. *Journal of Political Economy* 81: 950–61.

———. 1973b. The economics of overexploitation. *Science* 181: 630–34.

Conway, C.A. Global population structure of blue whales, *Balaenoptera musculus ssp.*, based on nuclear genetic variation. PhD thesis, University of California Davis. 2005.

Corkeron, P.J., and R.C. Connor. 1999. Why Do Baleen Whales Migrate? *Marine Mammal Science* 15: 1228–45.

Costa, D.P. 1993. The secret lives of marine mammals. *Oceanography* 6: 120–128.

Croll, D.A., and B.R. Tershy. 2002. Filter feeding, in *Encyclopedia of Marine Mammals*. Edited by W.F. Perrin, B. Wursig and J.G.M. Thewissen. San Diego: Academic Press.

Croll, D.A., et al. 1998. An integrated approach to the foraging ecology of marine birds and mammals. *Deep-Sea Research* II 45: 1353–71.

———. 2001. Effect of anthropogenic low-frequency noise on the foraging ecology of *Balaenoptera* whales. *Animal Conservation* 4: 13–27.

———. 2002. Only male fin whales sing loud songs. *Nature* 471: 809.

———. 2005. From wind to whales: trophic links in a coastal upwelling system. *Marine Ecology Progress Series* 289: 117–30.

Cummings, W.C., and P.O. Thompson. 1971. Underwater sounds from the blue whale, *Balaenoptera musculus*. *Journal of the Acoustical Society of America* 50: 1193–1198.

Deraniyagala, P.E.P. 1948. Some mystacetid whales from Ceylon. *Spolia Zeylanica* 25: 61–63.

Donovan, G.P. 1984. Blue whales off Peru, December 1982, with special reference to pygmy blue whales. *Report of the International Whaling Commission* 34: 473–76.

———. 1991. A review of IWC stock boundaries. Report of the International Whaling Commission, Special Issue 13: 39–68.

———. 2000. A note on the possible occurrence of pygmy blue whales (*Balaenoptera musculus brevicauda*) south of 60°S. International Whaling Commission paper SC/52/OS15.

Edds, P.L., and J.A.F. Macfarlane. 1987. Occurrence and general behaviour of balaenopterid cetaceans summering in the St. Lawrence estuary, Canada. *Canadian Journal of Zoology* 65: 1363–76.

Etnoyer, P., D. Canny, and B.R. Mate. 2006. Sea-surface temperature gradients across blue whale and sea turtle foraging trajectories off the Baja California Peninsula, Mexico. *Deep Sea Research II* 53: 340–58.

Fiedler, P.C., et al. 1998. Blue whale habitat and prey in the California Channel Islands. *Deep Sea Research II* 45: 1781–1801.

Findlay, K., et al. 1998. 1997/1998 IWC-Southern Ocean Whale and Ecosystem Research (IWC-SOWER) blue whale cruise, Chile. International Whaling Commission paper SC/50/Rep 2.

Gales, N.J., et al. 2005. Japan's whaling plan under scrutiny. *Nature* 435: 883–84.

Galletti Vernazzani, B., et al. 2006. Blue, sei and humpback whale sightings during 2006 field season in northwestern Isla de Chiloé, Chile. International Whaling Commission paper SC/58/SH17.

———. 2007. Status of blue whales off Isla de Chiloé, Chile, during 2007 field season. International Whaling Commission paper SC/59/SH1.

Garde, E., et al. 2007. Age-specific growth and remarkable longevity in narwhals (*Monodon monoceros*) from West Greenland as estimated by aspartic acid racemization. *Journal of Mammalogy* 88: 49–58.

Gauthier, J.M., C.D. Metcalfe and R. Sears. 1997. Chlorinated organic contaminants in blubber biopsies from northwestern Atlantic Balaenopterid whales summering in the Gulf of St. Lawrence. *Marine Environmental Research* 44: 201–23.

Gauthier, J.M., and R. Sears. 1999. Behavioral response of four species of balaenopterid whales to biopsy sampling. *Marine Mammal Science* 15: 85–101.

George, J.C., et al. 1999. Age and growth estimates of bowhead whales (*Balaena mysticetus*) via aspartic acid racemization. *Canadian Journal of Zoology* 77: 571–80.

Gill, P.C. 2002. A blue whale (*Balaenoptera musculus*) feeding ground in a southern Australia coastal upwelling zone. *Journal of Cetacean Research Management* 4: 179–87.

Gingerich, P.D. 2005. Cetacea, in *The Rise of Placental Mammals*, edited by K.D. Rose and J.D. Archibald. Baltimore: Johns Hopkins University Press.

Goldbogen, J.A., et al. 2006. Kinematics of foraging dives and lunge-feeding in fin whales. *Journal of Experimental Biology* 209: 1231–44.

Hammond, P.S., R. Sears and M. Bérubé. 1990. A note on the problems in estimating the number of blue whales in the Gulf of St. Lawrence from photo-identification data. *Report of the International Whaling Commission*, Special Issue 12: 141–42.

Hucke-Gaete, R., and B.R. Mate. 2005. Feeding season movements and fall migration to wintering areas for Chilean blue whales. Presentation to the 16th biennial conference on the Biology of Marine Mammals, San Diego, California, December 12–16, 2005.

Hucke-Gaete, R., F.A. Viddi and M.E. Bello. 2005. Blue whales off southern Chile: overview of research achievements and current conservation challenges. International Whaling Commission paper SC/57/SH5.

Hucke-Gaete, R., et al. 2004. Discovery of a blue whale feeding and nursing ground in southern Chile. *Proceedings of the Royal Society of London, Series B* 271: S170–3.

Ichihara, T. 1961. Blue whales in the waters around Kerguelen Island. *Norsk Hvalfangst-Tidende* 50: 1–20.

———. 1963. Identification of the pigmy blue whale in the Antarctic. *Norsk Hvalfangst-Tidende* 52: 128–30.

———. 1966. The pygmy blue whale, *Balaenoptera musculus brevicauda*, a new subspecies from the Antarctic. In *Whales, Dolphins, and Porpoises*, edited by K.S. Norris. Berkeley and Los Angeles: University of California Press.

Ivashin M.V., and A.A. Rovnin. 1967. Some results of the Soviet whale marking in the waters of the North Pacific. *Norsk Hvalfangst-Tidende* 57: 123–29.

Jonsgård, A. 1955. The stocks of blue whales in the North Atlantic Ocean and adjacent Arctic waters. *Norsk Hvalfangst-Tidende* 44: 505–19.

———. 1966. The distribution of Balaenopteridae in the North Atlantic Ocean. In *Whales, Dolphins, and Porpoises*, edited by K.S. Norris. Berkeley and Los Angeles: University of California Press.

Kato, H., et al. 2000. Body length distribution and sexual maturity of southern blue whales, with special reference to sub-species separation. International Whaling Commission paper SC/52/OS4.

————. 2002. Further developments on morphological and behavioural key for subspecies discrimination of southern blue whales, analyses from data through 1995–96 to 2001–02 SOWER cruises. International Whaling Commission paper SC/54/IA8.

Kingsley, M.C.S., and R.R. Reeves. 1998. Aerial surveys of cetaceans in the Gulf of St. Lawrence in 1995 and 1996. *Canadian Journal of Zoology* 76: 1529–50.

Kot, B.W. 2005. Ventral pouch flow rebound during engulfment-feeding in the rorqual whales. PhD thesis, University of California Los Angeles.

Laist, D.W., et al. 2001. Collisions between ships and whales. *Marine Mammal Science* 17: 35–75.

Lambertsen, R.H. 1983. Internal mechanism of rorqual feeding. *Journal of Mammalogy* 64: 76–88.

Lambertsen, R.H., N. Ulrich and J. Straley. 1995. Frontomandibular stay of Balaenopteridae: a mechanism for momentum recapture during feeding. *Journal of Mammalogy* 76(3): 877–99.

Lane, Lauriat Jr., 1991. Melville and the blue whale. *Melville Society Extracts* 87: 7–8.

Laurie, A.H. 1937. The age of female blue whales and the effect of whaling on the stock. *Discovery Reports* 15: 225–269.

Lavigne, D.M., et al. 1990. Lower critical temperatures of blue whales, *Balaenoptera musculus*. *Journal of Theoretical Biology* 144: 249–57.

LeDuc, R.G., et al. 2001. A preliminary Bayesian analysis of genetic data from southern hemisphere blue whales and its relevance to the question of degree of subspecies mixing at high latitudes. International Whaling Commission paper SC/53/IA25.

————. *In press*. Patterns of genetic variation in southern hemisphere blue whales, and the use of assignment test to detect mixing on the feeding grounds. *Journal of Cetacean Research and Management*.

Ljungblad, D.K., C.W. Clark and H. Shimada. 1998. A comparison of sounds attributed to pygmy blue whales (*Balaenoptera musculus brevicauda*) recorded south of the Madagascar Plateau and those attributed to "true" blue whales (*Balaenoptera musculus*) recorded off Antarctica. *Report of the International Whaling Commission* 48: 439–42.

Ljungblad, D.K., et al. 1997. Sounds attributed to blue whales recorded off the southwest coast of Australia in December 1995. *Report of the International Whaling Commission* 47: 435–39.

Mackintosh, N.A. 1929. The Discovery investigations: objects, equipment and methods—Part III: The marine biological station. *Discovery Reports* 1: 223–29.

Mackintosh, N.A. 1950. The work of the Discovery Committee. *Proceedings of the Royal Society of London, Series A* 202: 1–16.

Mackintosh, N.A., and J.F.G. Wheeler. 1929. Southern blue and fin whales. *Discovery Reports* 1: 257–540.

Mansfield, A.W. 1985. Status of the blue whale *Balaenoptera musculus*. *Canadian Field Naturalist* 99: 417–20.

Mate, B.R., B.A. Lagerquist and J. Calambokidis. 1999. The movements of North Pacific blue whales during the feeding season off southern California and southern fall migration. *Marine Mammal Science* 15: 1246–57.

Mate, B.R., R. Mesecar and B.A. Lagerquist. 2007. The evolution of satellite-monitored radio tags for large whales: one laboratory's experience. *Deep-Sea Research* II 54: 224–47.

McDonald, M.A., J.A. Hildebrand and S.C. Webb. 1995. Blue and fin whales observed on a seafloor array in the Northeast Pacific. *Journal of the Acoustical Society of America* 98: 712–21.

McDonald, M.A., S.L. Mesnick and J.A. Hildebrand. 2006. Biogeographic characterisation of blue whale song worldwide: using song to identify populations. *Journal of Cetacean Research Management* 8: 55–65.

McDonald, M.A., et al. 2001. The acoustic calls of blue whales off California with gender data. *Journal of the Acoustical Society of America* 109: 1728–35.

McHugh, J.L. 1974. The role and history of the International Whaling Commission. In *The Whale Problem*, edited by W.E. Schevill. Cambridge, MA: Harvard University Press.

Mellinger, D.K., and C.W. Clark. 2003. Blue whale sounds from the North Atlantic. *Journal of the Acoustical Society of America* 114: 1108–19.

Metcalfe, C., et al. 2004. Intra- and inter-species differences in persistent organic contaminants in the blubber of blue whales and humpback whales from the Gulf of St. Lawrence, Canada. *Marine Environmental Research* 57: 245–60.

Mitchell, E. 1974. Present status of northwest Atlantic fin and other whale stocks. In *The Whale Problem*, edited by W.E. Schevill. Cambridge, MA: Harvard University Press.

Mizroch, S.A. 1985. On the relationship between mortality rate and length in baleen whales. *Report of the International Whaling Commission* 35: 505–10.

Mizroch, S.A., and J.M. Breiwick. 1984. Variability in age at length and length at age in Antarctic fin, sei and minke whales. *Report of the International Whaling Commission* 34: 723–28.

Moore, M.J., et al. 1999. Relative abundance of large whales around South Georgia (1979–1998). *Marine Mammal Science* 15: 1287–1302.

Moore, S.E., et al. 2002. Blue whale habitat associations in the northwest Pacific: analysis of remotely-sensed data using a Geographic Information System. *Oceanography* 15: 20–5.

———. 2006. Listening for large whales in the offshore waters of Alaska. *BioScience* 56: 49–55.

Nieukirk, S.L., et al. 2004. Low-frequency whale and seismic airgun sounds recorded in the mid-Atlantic Ocean. *Journal of the Acoustical Society of America* 115: 1832–43.

Ohsumi, S. 1979. Interspecies relationships among some biological parameters in cetaceans and estimation of the natural mortality coefficient of the Southern Hemisphere minke whales. *Report of the International Whaling Commission* 29: 397–406.

Oleson, E.M., S.M. Wiggins and J.A. Hildebrand. *In press*. Temporal offset of blue whale call types on a southern California feeding ground. *Animal Behaviour*.

Oleson, E.M., et al. 2007a. Blue whale visual and acoustic encounter rates in the Southern California Bight. *Marine Mammal Science* 23: 574–97.

———. 2007b. Behavioral context of call production by eastern North Pacific blue whales. *Marine Ecology Progress Series* 330: 269–284.

Olsen E., and J. Sunde. 2002. Age determination of minke whales (*Balaenoptera acutorostrata*) using the aspartic acid racemization technique. *Sarsia* 87: 1–8.

Olsen, Ø. 1914–15. Hvaler og hvalfast i Sydafrika. Translated by Lilian Van Aarde, 1989. *Bergen Museums Aarbok* 5: 1–56.

Orton, L.S., and P.F. Brodie. 1987. Engulfing mechanics of fin whales. *Canadian Journal of Zoology* 65: 2898–2907.

Palacios, D.M. 1999. Blue whale (*Balaenoptera musculus*) occurrence off the Galápagos Islands, 1978–1995. *Journal of Cetacean Research Management* 1: 41–51.

Payne, R., and D. Webb. 1971. Orientation by means of long range acoustic signaling in baleen whales. *Annals of the New York Academy of Sciences* 188: 110–41.

Ramp, C., et al. 2006. Survival of adult blue whales *Balaenoptera musculus* in the Gulf of St. Lawrence, Canada. *Marine Ecology Progress Series* 319: 287–95.

Rankin, S., J. Barlow and K. Stafford. 2006. Blue whale (*Balaenoptera musculus*) sightings and recordings south of the Aleutian Islands. *Marine Mammal Science* 22: 708–13.

Rankin, S., et al. 2005. Vocalisations of Antarctic blue whales, *Balaenoptera musculus intermedia*, recorded during the 2001/2002 and 2002/2003 IWC/SOWER circumpolar cruises, Area V, Antarctica. *Journal of Cetacean Research and Management* 7: 13–20.

Rayner, G.W. 1940. Whale marking: progress and results to December 1939. *Discovery Reports* 19: 245–84.

Reilly, S.B., and V.G. Thayer. 1990. Blue whale (*Balaenoptera musculus*) distribution in the eastern tropical Pacific. *Marine Mammal Science* 6: 265–77.

Rice, D.W. 1966. Blue whales in the waters off Baja California. Report to the International Whaling Commission.

———. 1974. Whales and whale research in the eastern North Pacific in *Whales, Dolphins, and Porpoises*. Edited by K.S. Norris. Berkeley and Los Angeles: University of California Press.

———. 1986. Blue whale, in *Marine Mammals of Eastern North Pacific and Arctic Waters*. 2nd ed. Edited by Delphine Haley. Seattle: Pacific Search Press.

Rivers, J.A. 1997. Blue whale, *Balaenoptera musculus*, vocalizations from the waters off central California. *Marine Mammal Science* 13: 186–95.

Roman, J., and S.R. Palumbi. 2003. Whales before whaling in the North Atlantic. *Science* 301: 508–10.

Sears, R. 1987. The photographic identification of individual blue whales (*Balaenoptera musculus*) in the Sea of Cortez. *Cetus* 7: 14–17.

———. 2002. Blue whale *Balaenoptera musculus*, in *Encyclopedia of Marine Mammals*. Edited by W.F. Perrin, B. Wursig and J.G.M. Thewissen. San Diego: Academic Press.

Sears, R., and F. Larsen. 2002. Long range movements of a blue whale between the Gulf of St. Lawrence and West Greenland. *Marine Mammal Science* 18: 281–85.

Sears, R., et al. 1990. Photographic identification of the blue whale (*Balaenoptera musculus*) in the Gulf of St. Lawrence, Canada. *Report of the International Whaling Commission*, Special Issue 12: 335–42.

———. 1999. Gender related structure in blue whale (*Balaenoptera musculus*) pairs from the Gulf of St. Lawrence. Presentation to the 13th Biennial Conference on the Biology of Marine Mammals, Maui, Hawaii, November 28–December 3, 1999.

Sigurjónsson, J., and T. Gunnlaugsson. 1990. Recent trends in the abundance of blue and humpback whales off west and southwest Iceland, with a note on occurrence of other cetacean species. *Report of the International Whaling Commission* 40: 537–51.

Širović, A., et al. 2004. Seasonality of blue and fin whale calls and the influence of sea ice in the Western Antarctic Peninsula. *Deep-Sea Research* II 51: 2327–44.

Spilliaert, R., et al. 1991. Species hybridization between a female blue whale (*Balaenoptera musculus*) and a male fin whale (*B. physalus*): molecular and morphological documentation. *Journal of Heredity* 82: 269–74.

Stafford, K.M. 2003. Two types of blue whale calls recorded in the Gulf of Alaska. *Marine Mammal Science* 19: 682–93.

Stafford, K.M., C.G. Fox and D.S. Clark. 1998. Long-range acoustic detection and localization of blue whale calls in the Northeast Pacific Ocean. *Journal of the Acoustical Society of America* 104: 3616–25.

Stafford, K.M., and S.E. Moore. 2005. Atypical calling by a blue whale in the Gulf of Alaska. *Journal of the Acoustical Society of America* 117: 2724–27.

Stafford, K.M., S.E. Moore and C.G. Fox. 2005. Diel variation in blue whale calls recorded in the eastern tropical Pacific. *Animal Behaviour* 69: 951–58.

Stafford, K.M., S.L. Nieukirk and C.G. Fox. 1999a. An acoustic link between blue whales in the eastern tropical Pacific and the northeast Pacific. *Marine Mammal Science* 15: 1258–68.

———. 1999b. Low-frequency whale sounds recorded on hydrophones moored in the eastern tropical Pacific. *Journal of the Acoustical Society of America* 106: 3687–98.

———. 2001. Geographic and seasonal variation of blue whale calls in the North Pacific. *Journal of Cetacean Research Management* 3: 65–76.

Stafford, K.M., et al. 2004. Antarctic-type blue whale calls recorded at low latitudes in the Indian and eastern Pacific Oceans. *Deep Sea Research* I 51: 1337–46.

Stafford, K.M., et al. 2005. Location, location, location: acoustic evidence suggests three geographic stocks of "pygmy" blue whales in the Indian Ocean. Presentation to the 16th Biennial Conference on the Biology of Marine Mammals, San Diego, California, December 12–16, 2005.

Thompson, P.O. and W.A. Friedl. 1982. A long term study of low frequency sounds from several species of whales off Oahu, Hawaii. *Cetology* 45: 1–19.

Thompson, P.O., et al. 1996. Underwater sounds from blue whales, *Balaenoptera musculus*, in the Gulf of California, Mexico. *Marine Mammal Science* 12: 288–93.

True, F.W. 1899. On the nomenclature of the whalebone whales of the tenth edition of Linnæus's Systema Naturæ. *Proceedings of the U.S. National Museum* 21: 617–35.

Tyack, P.L., and C.W. Clark. 2002. Communication and acoustic behavior of dolphins and whales. In *Hearing by Whales and Dolphins*, edited by W.W.L. Au, A.N. Popper and R.R. Fay. New York: Springer-Verlag.

Watkins, W.A., et al. 2000. Seasonality and distribution of whale calls in the North Pacific. *Oceanography* 13: 62–67.

———. 2004. Twelve years of tracking 52-Hz whale calls from a unique source in the North Pacific. *Deep Sea Research* I 51: 1889–1901.

Wenzel, F., D.K. Mattila and P.J. Clapham. 1988. *Balaenoptera musculus* in the Gulf of Maine. *Marine Mammal Science* 4: 172–75.

Werth, A. 2000. Feeding in marine mammals. In *Feeding: Form, Function and Evolution in Tetrapod Vertebrates*, edited by K. Schwenk. New York: Academic Press.

———. 2001. How do mysticetes remove prey trapped in baleen? *Bulletin of the Museum of Comparative Zoology* 156: 189–203.

Wiggins, S.M., et al. 2005. Blue whale (*Balaenoptera musculus*) diel call patterns offshore of southern California. *Aquatic Mammals* 31: 161–68.

Williams, T.M., et al. 2000. Sink or swim: strategies for cost-efficient diving by marine mammals. *Science* 288: 133–36.

Yochem, P.K., and S. Leatherwood. 1985. Blue whale *Balaenoptera musculus* (Linnaeus, 1758). In *Handbook of Marine Mammals*, Volume 3, edited by S.H. Ridgeway and R.J. Harrison. London: Academic Press.

Zemsky, V.A., et al. 1995. Soviet Antarctic pelagic whaling after WW II: review of actual catch data. *Report of the International Whaling Commission* 45: 131–35.

Zenkovich, B.A. 1962. Sea mammals as observed by the round-the-world expedition of the Academy of Sciences of the USSR in 1957/58. *Norsk Hvalfangst-Tidende* 51: 198–210.

POPULAR MEDIA

Calambokidis, J. 1995. Blue whales off California. *Whalewatcher* 29(1): 3–7.

Chadwick, D. 2001. Evolution of the whale. *National Geographic* 200(5): 64–77.

Clark, Christopher W. 1994. Blue deep voices: insights from the Navy's Whales '93 Program. *Whalewatcher* 28(1): 6–11.

Conover, A. 1996. The object at hand. *Smithsonian* 27(7): 28.

Donahue, M.A., W.L. Perryman and J.W. Gilpatrick Jr. 1997. Measuring the blues below from the sky above. *Alolkoy* 9: 4.

Ellis, R. 1986. Hagiography of the whale. *Whalewatcher* 20(1): 11–15.

Komatsu, M. 2002b. The case for sustainable whaling. *Japan Echo* 29: 5.

Ohsumi, S. 1994. Why Antarctic blue whales do not recover. *Isana* 10.

Powell, M. 1998. The St. Lawrence Blues. *Equinox* 27(4): 47–53.

Reeves, R., and E. Mitchell. 1987. Hunting whales in the St. Lawrence. *The Beaver* 67(4): 35–40.

Ruud, J.T. 1956. The blue whale. *Scientific American* 195(6): 46–80.

Stein, D.L. 1992. The Prince of Whales. *Yankee* 56: 136.

Stevens, W.K. 1989. New survey raises concerns about recovery of blue whale. *New York Times*, June 20, 1989: C4.

Swartz, S. 1995. Scientific whaling and the IWC. *Whalewatcher* 29(1): 14–16.

Tarpy, C. 1979. Killer whale attack! *National Geographic* 155(4): 542–45.

Wong, K. 2003. The mammals that conquered the seas. *Scientific American*, Exclusive Online Issue, April 2003: 36–44.

GOVERNMENT AND NGO DOCUMENTS

Alling, A., E.M. Dorsey and J.C.D. Gordon. 1991. Blue whales (*Balaenoptera musculus*) off the Northeast coast of Sri Lanka: distribution, feeding and individual identification. UNEP Marine Mammal Technical Report 3: 247–258.

Barlow, J. 1995. The abundance of cetaceans in California waters. Part I: Ship surveys in summer and fall of 1991. *Fishery Bulletin* 93: 1–14.

Calambokidis, J. 2002. Final technical report: Underwater behavior of blue whales using a suction-cup attached Crittercam. Cascadia Research, Olympia, WA.

Calambokidis, J. 2003. Underwater behavior of blue whales examined with suction-cup attached tags. Cascadia Research, Olympia, WA.

Discovery Committee. 1937. Report on the progress of the Discovery Committee's investigations. Colonial Office, London.

Freeman, M.M.R. 1990. Political issues with regard to contemporary whaling, in *Who's Afraid of Compromise?* Edited by Simon Ward, Institute of Cetacean Research.

Gill, P.C., and M.G. Morrice. 2003. Blue whale research in the Bonney Upwelling, southeast Australia—current information. Technical paper 2003/1, School of Ecology and Environment, Deakin University.

Government of Japan. 2005. Plan for the second phase of the Japanese whale research program under special permit in the Antarctic (JARPA II)—monitoring of the Antarctic ecosystem and development of new management objectives for whale resources. International Whaling Commission paper SC/57/O1.

Kahn, B. 2003. Solor-Alor visual and acoustic cetacean surveys: interim report April-May 2003 survey period. Report to The Nature Conservancy and the SE Asia Center for Marine Protected Areas.

———. 2005. Indonesia Oceanic Cetacean Program Activity Report: April–June 2005, report to The Nature Conservancy and APEX Environmental.

Kahn, B., and J. Pet. 2002. Long-term visual and acoustic cetacean surveys in Komodo National Park, Indonesia 1999–2001: Management implications for large migratory marine life. In *Proceedings and Publications of the World Congress on Aquatic Protected Areas 2002*, Australian Society for Fish Biology.

Kahn, B., N.B. Wawandono and J. Subijanto. 2007. Positive identification of the rare pygmy Byde's whale (*Balaenoptera edeni*) with the assistance of genetic profiling. Report to The Nature Conservancy and APEX Environmental.

Reeves, R.R., et al. 1998. Recovery plan for the blue whale (*Balaenoptera musculus*). Office of Protected Resources, National Marine Fisheries Service, NOAA, Silver Spring, MD.

———. 2003. Dolphins, Whales and Porpoises: 2002–2010 Conservation Action Plan for the World's Cetaceans. Cetacean Specialist Group. IUCN, Gland, Switzerland, and Cambridge, UK.

Rice, D.W. 1992. The blue whales of the southeastern North Pacific Ocean. AFSC Quarterly Report, Oct-Nov-Dec 1992, Alaska Fisheries Science Center, Seattle.

Sears, R., and J. Calambokidis. 2002. COSEWIC assessment and update status report on the blue whale *Balaenoptera musculus* in Canada. Committee on the Status of Endangered Wildlife in Canada, Ottawa.

Stenson, G., et al. 2003. Ice entrapments of blue whales in southwest Newfoundland: 1868–1992, abstract in proceedings of the workshop on the development of research priorities for the northwest Atlantic blue whale population, November 20–21, 2003, Quebec City. Department of Fisheries and Oceans. Canadian Science Advisory Secretariat.

Index

Page numbers in italic type denote pages with photos, illustrations, or maps.

accidental deaths, 212, 264–65
Account of the Arctic Regions, An
   (Scoresby), 196
Acevedo-Gutiérrez, Alejandro, 159
acoustic research. *See also* vocalizations
   acoustic recording packages, 175
   adaptive beamforming, 173
   DIFAR sonobuoy, 187
   early, 164–65
   future avenues of, 191
   to identify subspecies, 231–33
   limitations of, 178
   measurement of sound levels, 166–67
   navy hydrophones used for, 171–73
   obstacles to, 165
   Payne's 1971 breakthrough, 167–71
   potential of, 173–75
   sound waves underwater, 162–63
   spectrograms, 176–77, *177*
Africa, 8, 24, 69, 107, 112, 183, 237
Aleutian Islands, 6, 133, 134, 135–37,
   136, 138, 140, 182
Ambulocetus natans, 243
American Museum of Natural History,
   17–18
Among Whales (Payne), 240
Ann Alexander (whaleship), 24
Annals of the New York Academy of Sciences,
   168
Antarctic blue whale (*Balaenoptera musculus
   intermedia*). *See also* blue whale
   classification of, 193–94, 200, 219–20

identification of, 227–33
migration of, 236–37
numbers of, 228
recovery of, 236
vocalizations of, 231–32
Antarctic convergence, 30, 70
Antarctica
   beginning of commercial whaling in,
      32–33
   description of whaling in, 50–53
   development of pelagic whaling in,
      36–38
   feeding grounds in, 226–27
   golden age of whaling in, 39–45
   identified as location for whaling, 30–31
   number of blue whales in, 12–13, 259,
      261, 271–74
   postwar decline of whaling in, 49, 61,
      63–65, 78, 80–81
   research in, 68–69, 69–72
Arabian Nights Entertainments, 13
archaeocetes, 242–43
Arctic Ocean, 23, 30, 31, 267
Argentina, 74, 103, 172, 196, 198
Argos system, 144, 145–46
Aristotle, 195
ARPs, 175
Ash, Christopher, 86, 87
Australia
   blue whales off, 7
   endangered species status in, 260
   feeding areas off, *204*

migration to and from, 209–10, 237
research in, 203–7
supports regulation, 80
threat from shipping in, 265
Azores, 96, 106, *107*, 111–12, 183

Baffin Island, 107
Baja California, 6, 96, 109, 132–33, *136*,
   139, 140, 147
Balaenoptera (genus), 5
Balaenopteridae (family), 4
baleen
   of blue whale discarded, 55
   characteristics of, 196–97, 250
   described, 3–4
   uses of, 22, 34
baleen whales
   classification of, 3–4
   evolution of, 243–44
   feeding strategies of, 250–51, 254
   longevity of, 12
   meat of, 49, 53
   migration of, 108–9
   vocalizations of, 191
Baleine Bleu (research boat), 94
Barlow, Jay, 124, 126, 137
Barnes, R.H., 13
Basques, 2–3
beaked whale, 3, 270
Bechmann, Hans, 38, 39–42, 44, 61, 63,
   65
Bechmann, Johan Ludvig, 39
belugas, 3, 267, 269
Bendlin, Andrea, 93–95, 104
Benko, Fred, 130–31
Berchok, Catherine, 173–74, 176
Bermuda, 110, 183
bioacoustic probe, 156–58, 160, 188
biopsies, 114–15, 128–29, 149–50,
   222–23. *See also* research
birthing, 8, 148–49
blowholes
   chemical receptors around, 161
   in different subspecies, 231
   in different whale groups, 4
blue whale (*Balaenoptera musculus*). *See also*
   Antarctic blue whale; pygmy blue whale
   as conservation icon, 88–91
   death throes of, 43, 86–87
   dimensions of, 2
   exaggerations concerning, 1–2, 9, 16,
      167, 239–40, 250, 251
   historic description of, 14–15

lack of knowledge about, 2–3, 67, 90,
   215
length of, 239–41
longevity of, 10–12, 263
models of, 17–18
numbers of, 12–13
origin of name, 19–20
populations of, 185, 224, 227, 260
sense of hearing, 163–64
sense of sight, 161
sense of smell, 161
sense of taste, 162
size of, 241–42, 249
speed of, 147
subspecies of, 193–202
threat from noise, 270–71
weight of, 241
Blue Whale, The (Small), 54, 90–91, 199,
   239, 259
blue whale unit (BWU), 75–76, 77
Blyth, Edward, 196, 202
bomb-lances, 26, 131
Bonney Upwelling, *204*, 204–5, 207, 237
bowhead whale (*Balaena mysticetus*), 4, 78,
   195
   baleen of, 250
   feeding strategy of, 250
   longevity of, 12
   numbers of, 23, 262
   tagging of, 144
   whaling of, 23, 25
B-probe. *See* bioacoustic probe
Branch, Trevor, 233–36, 238, 272–73
Brendan, Saint, 13
British Columbia, 103, 121, 133, 135–37
broad-nosed whale, 15
Bryde's whale (*Balaenoptera edeni*), 5, 277
buoy boats, 51, 77
Bureau of Whaling Statistics, 76, 83
Burmeister, Hermann, 196–97, 201
BWU, 75–76, 77

Calambokidis, John, 121–24, 126–31,
   129, 134–37, 139–40, 143, 150, 182,
   186, 190, 219, 270
   bioacoustic probe, 158, 187–88
   Crittercam research, 153–56
   on need for ongoing research, 141
   satellite tagging project, 138, 144–45,
      147–48, 179
California
   migration from, 147, 179
   population of blues off, 6, 7, 121–41

whale-watching off, 266
whaling off, 131–32
Call to the South (McLaughlin), 53
calving, 8, 10, 148–49
Canada
    endangered species status in, 259–60
    opposition to regulation, 80
    quits IWC, 79
    as whaling nation, 108
Carson, Rachel, 87
Cascadia Research Collective, 123, 134,
    137, 140
catcher boats, 40, *41*
    in Antarctic fleet, 36–38, 51
    life aboard, 58–62, *59*, 86–87
    speed of, 40, 60
    during Second World War, 48
Centro Ballena Azul (Blue Whale Cen-
    tre), 216, 218
Centro de ConservaciÛn Cetacea, 218,
    220
Chadwick, Douglas, 244
Chile
    feeding areas off, 6–7, 217
    opposition to regulation, 74, 79, 80
    research in, 213–20
    subspecies of blue whale off, 219–20,
        224, 232–33, 236, 238
    whale-watching off, 266
    whaling by Europeans off, 213–14
    as whaling nation, 74, 79, 80
Christian Salvesen and Company, 41, 50,
    55, 57, 62, 65
Clark, Chris, 171–73, 183, 184, 270
classification. *See* taxonomy
climate change, 138, 269–70
Committee of Three, 80, 271
CompaÒÌa Argentia de Pesca, 32, 33
Condor Express (whale-watching boat),
    130–31
Congreve, William, 26
conservation. *See* whaling, regulation of
Conway, Carole, 30, 221–27, 229, 234,
    237
Cook, James, 30
Cope, Edward Drinker, 197
corpora albicantia, 11, 235
Costa Rica Dome, 6, *136*, *179*
    acoustic research in, 178–80
    as calving ground, 148–50
    migration to, 139–40, 147
    mixing of populations in, 224–26
Crittercam, 152–56

Croll, Don, 159, 245–49, 270
Cummings, William, 166, 167, 170, 180,
    215, 233

Davis Strait, 6, *107*, 107, 110, 183, 239
DDT, 267, 268–69
de la Roche, Anthony, 30
deep isothermic layer, 162–63
deep sound channel, 163, 168, 170, 175,
    190
Dewhurst, Henry William, 15–16
dinosaurs, 2, 241, 242
Discovery (research ship), 68
Discovery II (research ship), 69–70
Discovery Committee, 68–73, 143, 234
Discovery Reports, 69
diving. *See also* feeding
    characteristics of, 127–28
    negative buoyancy, 158
    theoretical aerobic dive limit, 159
dolphins
    classification of, 3
    in mythology, 13
    vocalizations of, 168, 189
Don Ernesto (ship), 39
Doniol-Valcroze, Thomas, 113, 114,
    115, 117, 118

echo sounder, 127
Ellis, Richard, 28, 33, 35, 78, 88–89, 259
Encyclopaedia Britannica, 19
Encyclopedia of Marine Mammals, 239
endangered species status, 81, 130, 140,
    259–61
Essex (whaleship), 24
evolution, 242–44
extinction of animal species, 44–45

factory ships, *39*, 52
    in Antarctic fleet, 36–38, 51
    during Second World War, 48
feeding. *See also* diving; krill; lunging
    adaptability of blues, 269
    anatomy, 250–51, 253
    biomechanics, 252–57
    at depth, 156, 158, 254
    feces, 204
    food requirements, 249–50
    grounds, 6, 6–7, 244–49
    as solitary activity, 156
    strategies, 147–48, 155, 205–6, 211,
        248–49, 251, 255–56, 257–58
    vocalization and, 189

fin whale *(Balaenoptera physalus)*
classification of, 5
feeding biomechanics of, 257
numbers of, 262
protection of, 80
in St. Lawrence, 99, 269
scientific whaling of, 277
tagging of, 144
threat from noise, 270–71
vocalizations of, 89–90, 167, 168,
169–70, 190, 191
whaling of, 49, 63, 78
Finnmark, 29, 31, 32, 107, 112
Firth of Forth, 14, 195
Flemish Cap, *107*, 110, 111
Føyn, Svend, 19–20, 25–29, 31, 41
Francis, John, 153
Fraser, F.C., 199
Fraser, Gibbie, 57–61, *59*, 62, 86–87

Galletti, Bárbara, 218–19, 234
Geneva Convention for the Regulation
of Whaling, 73–74
Germany, 48, 49, 74, 75
Gibson, John, 97
Gill, Peter, 203–7, *207*, 234, 265
Gilpatrick, Jim, 134, 135, 200–201
Gore, Al, 171
Grand Banks, *107*, 110, 111, 171–72
Greeneridge Sciences, 156
Greenland, 104, 106, 107, 110
Greenland whale. *See* bowhead
Greenpeace, 87
grey whale *(Eschrichtius robustus)*
classification of, 4
curiosity of, 90
decline of, 78
feeding strategy of, 250–51
migration of, 7, 108–9, 148, 149
numbers of, 262
photo-identification of, 103
protection of, 74
recovery of, 140
Guinea, 112
Gulf of Alaska, 131–34, 137–38, 140,
181–83, 263
Gulf of Maine, 110
Gulf of the Farallones, 123, 131, 133,
136, 248, 249

Handbook of Marine Mammals, 240
harpoons, 22, 23, 212–13
harpoon cannons, *41*

evolution of, 61–62
invention of, 19–20, 26, 27–28
operation of, 42–43
Hawke's Harbour, 108
Hector Whaling, 62
History of Modern Whaling (Tønnessen
and Johnsen), 43
Hucke-Gaete, Rodrigo, 215–18, 219,
220, 234
Hudson's Bay Company, 68
humpback whale *(Megaptera
novaeangliae)*
calving grounds of, 148
characteristics of, 201
classification of, 5
contaminants in, 268
feeding pattern of, 114, 206
lunging by, 254
migration of, 7, 105, 109, 149
numbers of, 262
in St. Lawrence, 99, 269
scientific whaling of, 277
tagging of, 144, 145
vocalizations of, 88–90, 117, 178, 186,
190
whaling of, 23, 33, 203
hybrid species, 6

Iceland
blue whales off, 96, 106, 107, 183,
266–67
opposition to regulation, 79, 108, 274
photo-identification in, 111–12
renewed whaling by, 277–78
Ichihara, Tadayoshi, 198–200, 228–29,
235
illegal whaling, 82–84, 259
Indian Ocean
migration in, 209–10
populations of blue whales in, 226–27,
232–33, 237–38
whaling in, 81, 198, 200, 238
Indonesia
aboriginal whaling in, 79, 212–13
blue whale areas of, *209*
conservation in, 211–13
mythology of whales in, 13–14
research in, 208–11
threats to blues in, 212–13
Integrated Undersea Surveillance
System, 165, 171, 173, 183
International Convention for the
Regulation of Whaling, 74, 78

International Whaling Commission. *See also* whaling, regulation of
  database of, 234, 236, 239
  definition of "management stock" by, 106, 140
  permits scientific whaling, 276–77
  politics of, 78–84, 274–75
  research by, 104, 132
  Scientific Committee of, 78, 202, 229, 275, 278
IUSS. *See* Integrated Undersea Surveillance System
IWC. *See* International Whaling Commission

Japan
  consumption of whale meat in, 53, 277
  evades regulations, 214
  opposition to regulation, 74, 75, 80–81, 199–200, 274–78
  research by, 202, 203, 241, 276–77
  as whaling nation, 5, 29, 48, 49, 64, 202
Jason (ship), 31
John H. Martin (research ship), 185–88
Johnsen, A.O., 43
Jonah, Book of, 13

Kahn, Benjamin, 207–11, 234, 237
Kamchatka Peninsula, 6, 132, 134, 140
Kato, Hidehiro, 230–31
Katona, Steven, 88, 89–90, 98, 103
Kennedy, Ted, 171
killer whale. *See* orca
Komatsu, Masayuki, 275, 276
Kot, Brian, 254–58
krill. *See also* feeding; upwelling zones
  characteristics of, 4, 244, 247–48
  climate change and, 138, 269–70
  evolution of, 243
  life cycle of, 246–47
  locating patches of, 127
  as primary food source, 69

Labrador, 107, 108
Lambertsen, Richard, 252–53, 257
Lancing (ship), 38
Lane, Lauriat, Jr., 19
Larsen, Carl Anton, 31–33, 36–38, 40
League of Nations, 73
LeDuc, Rick, 229–30, 232–33, 234
Les Escoumins, 95, 99, 265
Lever Brothers, 35

line-transect method, 124, 125, 174, 271
Linnaeus, Carolus, 15, 194–96, 197
lunging, 158, 159–60, 254–55. *See also* feeding

Mackintosh, Neil, 69, 72, 73
March of the Penguins, 153, 154
Marine Mammals of the North-western Coast of North America, The (Scammon), 197
Marine Mammals of the World (Rice), 201–2
mark-recapture method, 124–26, 174, 267
Marshall, Greg, 152–53
Matamek Research Station, 97–98
Mate, Bruce
  on difficulties of whale research, 151–52
  on feeding strategies, 248–49
  future avenues of research, 160
  on importance of research, 140
  on limitations of photo-ID, 145
  on migration patterns, 148
  satellite tagging project, 138–40, 144–45, 147, 149–50, 179, 218
Mathisen, Alf, 61, 63
mating, 8–9, 118–19, 170
Mauritania, 112
McArthur (research ship), 124
McHugh, J.L., 81–82
McLaughlin, W.R.D., 53, 87
McVay, Scott, 88
Meiklejohn, James, 50–51, 53, 56–57, 65
Mellinger, David, 183, 184
Melville, Herman, 18–19, 20, 24
MICS. *See* Mingan Island Cetacean Study
migration
  in Antarctica, 71–72
  changes in patterns of, 137–38
  convergence of different populations, 180–82
  in eastern North Atlantic, 111–12
  in eastern North Pacific, 132–34, 135–41, 179–80
  mystery surrounding, 7–8, 108–9
  variations in patterns of, 148
  in western North Atlantic, 109–11
Mingan Island Cetacean Study, 98–105, 108–19, 173, 268
Mingan Islands, 99, 99, 105–6, 263
minke whale (*Balaenoptera acutorostrata* and *B. bonaerensis*)
  classification of, 5

competing with blues for food, 275–76
feeding, 257
longevity of, 12
lunging by, 254–55
numbers of, 275–76
in St. Lawrence, 94, 99
scientific whaling of, 277
whaling of, 64, 202
Moby-Dick (Melville), 18, 24, 218
Monterey Bay, 131, *136*, 248, 266
Moore, Sue, 182
Morrice, Margie, 204–6, 207, 234, 265
Moss Landing Marine Laboratories, 185
Mowat, Farley, 67, 199
Musculus (research ship), 216
mysticetes, 3–4, 15. *See also* baleen whales
mythology, 13–14

Nantucket, 23
Nantucket sleigh ride, 26
narwhal, 12
National Geographic Society, 153
Nature (journal), 273
Netherlands
    exits Antarctic whaling, 64, 81
    opposition to regulation, 79
    as whaling nation, 23, 49, 78
New Bedford Whaling Museum, 264
New England, 23, 88, 98, 104, 106, 110
New Sevilla (factory ship), 47–48, 53
Newfoundland, 102, 107, 111, 172, 263
Newton, George, 17
Nieukirk, Sharon, 183–84
Nishiwaki, Masaharu, 201, 241
North Atlantic
    blues crossing, 112–13
    populations of blues in, 106–8, *107*,
        110, 183–85, 261
    whales of eastern, 111–12
    whaling in, 107–8
North Pacific. *See also* Pacific
    accidental deaths in eastern, 265
    blues in eastern, 126, *136*
    blues in western, 140
    populations of blues in, 133–34,
        135–41, 261
    whaling in, 131–32
Norway
    consumption of whale meat in, 55
    opposition to regulation, 79, 274
    supports regulation, 74, 80
    whaling in, 28–29
    as whaling nation, 29, 32, 48, 64, 214

Norwegian-Canadian Whaling
    Company, 108
Nova Scotia, 110

odontocetes, 3, 4. *See also toothed whales*
Ohsumi, Seiji, 276
Oleson, Erin, 185–91
Olympic Challenger (factory ship), 82–83
Onassis, Aristotle, 82
orca (*Orcinus orca*)
    attacks on blues by, 263–64
    classification of, 3
    difference in size between sexes, 4
    photo-identification of, 103
Ostend Whale, 16, *16*
Oxford English Dictionary, 19

Pacific. *See also* North Pacific
    hydrophones in eastern tropical, *179*
    mixing of populations in eastern
        tropical, 180–81, 224–26
    research in southeastern, 215–20
Pakicetus, 242
Papua New Guinea, 208
Paradise Lost (Milton), 13
Payne, Roger
    Among Whales, 240
    first encounter with blue, 2
    landmark paper on acoustics by, 89,
        168–71, 186
    recordings by, 88
    vindication of, 172–73
    work on photo-identification, 103,
        168
PCBs, 267, 268, 269
Perryman, Wayne, 134
Perth Canyon, *204*, 206
Peru, 6, 79, 80, 83, 229, 233
Peterson, Alex, 47–48, 53
photo-identification
    in Australia, 206
    off California, 124, 126, 127, 128
    in eastern North Pacific, 139–40
    invention of, 168
    limitations of, 145
    in St. Lawrence, 93, 95, 102–5
    in southeast Pacific, 218, 219
phytoplankton, 139, 161, 245, 246
piracy. *See* illegal whaling
Pliny the Elder, 13, 195
porpoises, 3
"Prince of Whales," 17
pygmy blue whale (*Balaenoptera musculus*

brevicauda). See also blue whale
in Antarctica, 232, 234–36
off Australia, 194, 203–7, 232, 237
characteristics of, 198
off Chile, 219–20
controversy over classification of,
    193–94, 199–202
identification of, 227–33
length of, 239
Myrbj̈nner and, 199, 235
numbers of, 228, 261
populations of, 232–33, 237–38
vocalizations of, 231–32
whaling of, 81, 203
pygmy Bryde's whale, 213
pygmy right whale, 4, 194
pygmy sperm whale, 194

Ramp, Christian, 115
Red List, 140
reproduction. *See also* calving
breeding grounds, 7, 148, 170
fertility, 266–67
gestation period, 9
interbreeding, 224–27
rate of, 9, 222, 272
sexual organs, 9, 235
size of fetus, 8
research. *See also* acoustic research;
biopsies; photo-identification; tagging
aerial photogrammetry, 134, 200
aerial surveys, 205–6, 216, 218
beginnings of, 67–73
Crittercam, 152–56
digital video, 256
estimating age, 11–12
estimating length, 134–35
estimating population size, 124–25
field notes, 101–2, 128
future avenues of, 160
genetic, 221–27, 229–30
obstacles to, 150–52
overview of, 7
video, 230–31, 254
Resolution (ship), 30
Rice, Dale, 132–33, 137, 201–2, 240–41,
    250
right whale (*Eubalaena* spp.)
classification of, 4
feeding strategy of, 250
migration of, 112–13
near-extinction of, 78
numbers of, 262–63

photo-identification of, 103
protection of, 74
in St. Lawrence, 98
tagging of, 144
tangled in fishing gear, 265
vocalizations of, 191
whaling of, 22, 23, 31, 203
rocket harpoon, 27
Roys, Thomas Welcome, 27, 28
Russia, 274. *See also* Soviet Union

St. Lawrence (Gulf and estuary)
blues injured by fishing gear in, 265
collisions with ships in, 264–65
map of, *99*
migration from, 109–11, 171–72
pollution in, 267–68
reproduction in, 266–69
research in, 93–102, 104–6, 108,
    113–14, 254–56
whale-watching in, 266
Salvesen. *See* Christian Salvesen and
Company
Salvesen Ex-Whaler's Club, 65
Santa Barbara Channel, 131, *136*, 154,
    248, 249, 265
Scammon, Charles, 26, 131, 197
Schevill, William, 167
Scoresby, William, 196, 197
Scotland, 14–15, 41, 57–59
Scott, Robert Falcon, 68
Sea of Cortez, 96, 104, *136*, 147, 149, 155
sealing, 25, 31
Sears, Richard, 7, 93–106, 97, 108–19,
    143, 171–72, 173–74, 183, 190, 257
on difficulties of whale research,
    151–52
on external threats, 264–66
on pygmy blue whale, 201
research on fertility and contaminants,
    266–69
sei whale (*Balaenoptera borealis*), 5, 64, 277
Sept-Œles, 108
sexes, 4, 114–15
Shackleton, Ernest, 68
Shetland Islands, 57–58
Sibbald, Sir Robert, 14–15
Sibbald's rorqual, 18
Siegesbeck, Johann, 195
Silent Spring (Carson), 87
Sir James Clark Ross (ship), 36–37, 42,
    44
sleeping, 155

Small, George, 54, 64, 80, 85, 90–91, 195, 199, 239, 259
Smithsonian Institution, 17, 197
social behaviour
  mating, 8–9, 118–19, 170
  pairs, 114, 115–17, 205–6
  "range herds," 170
  solitary habits, 113–14
  trios, 117–18
SOFAR channel, 163, 168, 170, 175, 190
Songs of the Humpback Whale (Payne), 88
South Africa, 8, 49, 69
South Georgia
  marine mammal survey in, 96
  overwintering in, 55–56
  research station on, 68–69
  shore stations on, 29–33, 81
South Korea, 80, 277
Southern Broom (catcher boat), 59–60
Southern Ocean. *See* Antarctica
Southern Ocean Whale and Ecosystem Research program, 202–3, 213, 215–16, 227, 228, 230, 231–32, 272, 277
Southern Soldier (catcher boat), 61
Southern Venturer (factory ship), 50–51, 56, 62, 65
Southwest Fisheries Science Center, 124, 134, 223, 229
Soviet Union. *See also* Russia
  illegal whaling by, 83–84, 200, 203, 206, 238
  opposition to regulation, 74, 77, 80, 81
  as whaling nation, 49, 64, 78
SOWER. *See* Southern Ocean Whale and Ecosystem Research program
Spain, 80
sperm oil, 24, 34
sperm whale (*Physeter macrocephalus*)
  classification of, 3
  difference in size between sexes, 4
  in Indonesia, 208, 213
  in *Moby-Dick*, 19
  numbers of, 262
  ramming ships, 43
  scientific whaling of, 277
  tagging of, 210–11
  towing whaleboats, 26
  whaling of, 14, 23–24, 203, 213, 262
spermaceti, 24
Sri Lanka, 7, 8, 194, 202, 232, 237–38
Stafford, Kate, 178–84, 191, 224, 232–33, 234

stern slipway, 38, 39
Strait of Belle Isle, 99, 107
sulphur-bottom whale, 18, 27, 131, 197
Swedish Antarctic Expedition, 32
Systema Naturae (Linnaeus), 195

tagging
  bioacoustic probe, 156–58, 160, 188
  early days of, 70–72
  GPS technology and, 160
  in Indonesian waters, 214–15
  in North Pacific, 132
  pop-up archival tags, 210–11
  recoverable data loggers, 152
  satellite, 138, 144–50, 152, 174, 218
  time-depth recorders, 159–60
taxonomy, 3–5, 14–15, 193–202, 219–20
telemetry. *See* tagging, satellite
Tershy, Bernie, 159, 245, 248
thermocline, 162–63
Thompson, John, 58
Thompson, Paul, 166, 167, 180, 215, 233
Thoris (catcher boat), 63
Tonga, 79
tongue, 53, 251–52
Tønnessen, J.N., 43
toothed whales
  characteristics of, 3–4, 161
  evolution of, 243
  vocalizations of, 164, 168
True, Frederick, 197–98

Unilever, 35
United Kingdom
  quits Antarctic whaling, 64, 81
  supports regulation, 74, 80
  as whaling nation, 23, 32, 49
United States, 49, 80, 82, 274
United States Navy, 89, 164–66, 167, 171, 232, 270
upwelling zones. *See also* krill
  Antarctic convergence, 30
  Bonney Upwelling, 204, 204–5, 207, 237
  causes of, 204, 245–47
  off Chile, 218, 219
  coastal, 245, 246, 248
  Costa Rica Dome, 139
  in Indonesia, 209
  in St. Lawrence, 99–100
  in tropical Pacific, 225

Vancouver Island, 132, 133

Vesterlide (whaling ship), 214
Villiers, A.J., 37, 42–43, 44, 73, 85–86
vocalizations. *See also* acoustic research
  "bilingual," 182, 226
  characteristics of, 175–76, 177–78,
    188–89
  dialects, 178–85
  of eastern North Pacific, 188
  first recording of, 166, 167, 215
  inspiring researchers, 88–89, 271
  length of, 190
  loudness of, 167
  of North Atlantic, 183–85
  purpose of, 185–86, 190–91
  recording of, 152, 153, 157
  role of gender in, 186–87, 188–89, 190
  source of, 10, 164
  unique to individuals, 173, 174
  of western North Pacific, 181
  whale's position during, 189–90

Watkins, William, 167
Webb, Douglas, 89, 168, 170
Werth, Alex, 253–54
whale claw, 38, *39*
whale meat
  consumption of, 22, 53–55, 277
  as pet food, 54
  postwar market for, 49
  shrinking market for, 34
whale oil
  amount produced by blue whale, 24,
    34
  early trade in, 22, 23
  extraction of, 53
  falling price of, 25, 64, 73
  postwar demand for, 49, 84
  used for soap and margarine, 34–36,
    67
  used in nitroglycerine, 35

Whaler's Eye (Ash), 86
Whales '93, 171
whale-watching, 90, 94, 110, 129–31,
  265–66
whaling, 21–87. *See also* catcher boats;
  factory ships; illegal whaling;
  International Whaling Commission;
  and individual locations
  aboriginal, 21, 79, 212–13
  aircraft used in, 62, 77
  becomes international business, 23
  blues become target, 29
  butchering and rendering, 51–53, 52,
    85
  crewmen's pay, 41
  crewmen's viewpoint, 84–87
  deaths of crewmen, 40
  electrocution used in, 61–62
  in First World War, 34–35
  heyday in 1930s, 44
  moratorium on, 82, 104, 274
  numbers killed, 44, 45, 48, 49, 60,
    72–73
  obstacles to, 24, 26–27, 28, 38
  postwar decline of, 49
  products derived from, 55
  regulation of, 73–82, 198–200, 236,
    259
  scientific, 276–77
  in Second World War, 47–49, 75
  sonar used in, 62–63, 77
  s150team-driven whaleships, 25–26
  technological advances in, 36, 38
  whales fighting back, 43–44
Whaling in the Frozen South (Villiers), 85
Wheeler, John, 69
William Scoresby (research ship), 70, 71
Woods Hole Oceanographic Institution,
  97
World Conservation Union, 140, 260–61